Tectonics of the South China Block
Interpreting the Rock Record

Zheng-Xiang Li   Han-Lin Chen   Xian-Hua Li
Feng-Qi Zhang

Science Press
Beijing

Brief information

South China experienced an unusually complex tectonic history, and featurs a series of globally significant, and well-preserved geological records. These include records of the assembly and breakup of the Neoproterozoic supercontinent Rodinia, Rodinia- and Pangea-age mantle plume events, Neoproterozoic glacial events, early complex life, large intraplate orogenic and magmatic events, extinction events around the P-T boundary, continental subduction and exhumation, and record of the transition from an Andean-type active margin to the present-day Western Pacific-type active margin. These features make South China one of the few natural laboratories for studying fundamental geoscience problems and testing various theories and hypotheses.

Professor Zheng-Xiang Li and his multinational collaborative team have been working on the tectonic evolution of the South China Block for over 25 years. Their work challenged some traditional views, yet some of their new interpretations remain controversial amongst contemporary researchers. This book provides readers a summary of their views on the tectonic evolution of South China, and evidence that their interpretations were based on. It also provides a well illustrated eight-day field program in which the authors attempt to unravel the tectonic history of eastern South China through the examination of a series of carefully selected field traverses and outcrops. This latter part of the book can also serve as materials for field workshops or short courses on tectonic analysis using multidisciplinary field observations and analytical results.

The book is designed for researchers of all levels and senior geoscience students.

Responsible Editor: Han Peng

Copyright© 2014 by Science Press
Published by Science Press
16 Donghuangchenggen North Street
Beijing 100717, China

Printed in Beijing

ISBN 978-7-03-040486-2

Preface

South China is the cradle of Chinese geology, where almost a century ago scientists both from within China and the west started ground breaking geological mapping, leading to the first understanding of the geological history of the region. Systematic 1:200,000 geological mapping was completed almost half a century later. The quality of that monumental work is such that those maps still serve as the solid foundation for geological research and resource exploration in the region today.

South China experienced an unusually complex tectonic history, featuring repeated orogenic and magmatic reworking since Neoproterozoic time. Partly due to this complex history, tectonic interpretations for the region have remained controversial. Nonetheless, some unique features found in the geological record of South China have brought the region to the global geoscience community's attention for the past decades, which in-turn enhanced the geoscience research in the region, and led to significant advances in fundamental geoscientific knowledge. These features include a possible record of supercontinent assembly and breakup events (timing, configuration, and mechanisms), repeated mantle plume events (possible superplume events, supercontinent-superplume coupling and plume generation — geodynamics), Neoproterozoic glacial record (the Snowball Earth hypothesis), a superb record of early complex life, large intraplate orogenic and magmatic events (far-field effects of continental collision, flat-subduction and foundering), P-T boundary and extinction events, and continental subduction and exhumation. These features thus make South China one of the few natural laboratories for the study of a range of fundamental geoscience problems and a testing ground for a range of globally-significant theories.

This book is designed to serve as a starting point for people wanting to better understand the tectonic evolution of South China, with a particular emphasis on eastern South China. It starts with a tectonic overview (Part 1, mainly authored by Zheng-Xiang Li) that summarizes some of the recent advances in tectonic research in the region, from its basement composition, to its assembly in the late Precambrian, and major tectonic events in the Phanerozoic. Local to regional observations are interpreted in the context of the broader-scale tectonic background.

Part 2 of the book, contributed by all authors, is an attempt to examine some of the tectonic theories and hypotheses, as summarized in Part 1, through an eight-day traverse in eastern South China, including information about selected outcrops and related analytical results. Readers can refer back to Part 1 for tectonic interpretations and key references for alternative tectonic models. This part is also designed that it can serve as a guide book for either organized field workshops or self-guided geological tours. Daily exercises at the back of each day's program are designed for training research students to construct time-space diagrams and use them for terrane and regional tectonic analyses. A Chinese version of Part 2 is provided as an appendix.

We are grateful to everyone who collaborated with us over the past decades on South China research, or assisted with fieldtrips and data analyses. In particular, we wish to thank Professors Wuxian Li, Jian Wang, Shihong Zhang, Yigang Xu, and Mr Chaomin Bao for their support and cooperation, and Dr Nick Timms for proof-reading the manuscript.

Authors
17 October, 2013

Contents

PART 1

A TECTONIC OVERVIEW OF THE SOUTH CHINA BLOCK

The South China Block (SCB) is bounded by the Qinling-Dabie-Sulu orogenic belt to the north, the Longmenshan Fault to the northwest, the Red River Fault to the southwest, and the continental slope of the East and the South China seas to the southeast, with a possible extension to the Korean Peninsula (Figure 1.1). The northwestern portion of the SCB is widely

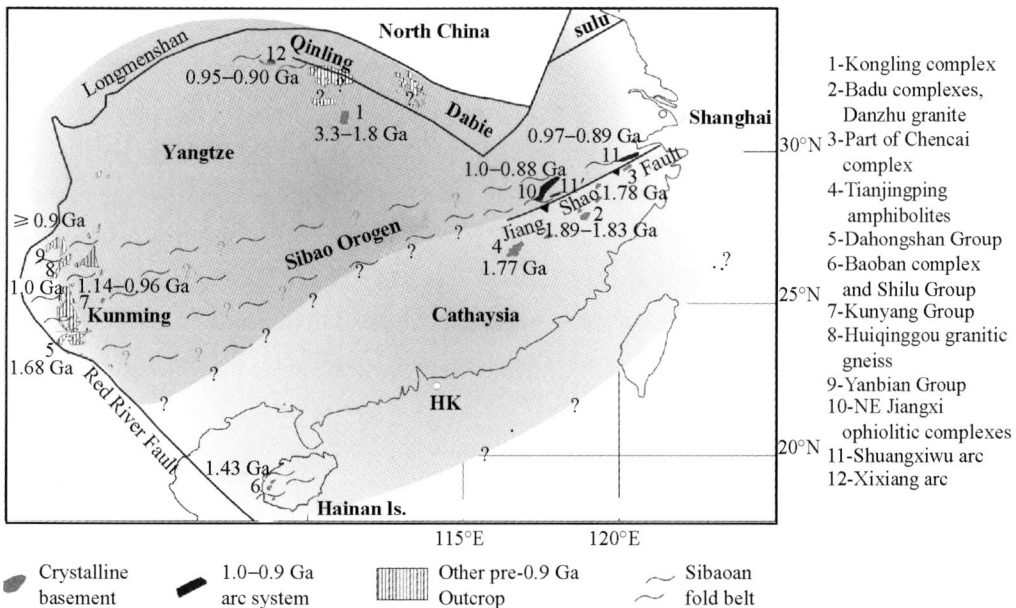

Figure 1.1   Tectonic framework of the South China Block, emphasizing pre-0.9 Ga (900 Ma) records (modified after Li Z X et al., 2007)

Sources for precisely dated rocks are: ① the Kongling Complex (Jiao et al., 2009; Qiu et al., 2000; Zhang et al., 2006; Gao et al., 2011), ② Badu magmatic and metamorphic complexes (Li and Li, 2007; Xiang et al., 2008; Yu et al., 2009), ③ central Chencai Complex (Li Z X et al., 2010), ④ the Tianjingping amphibolites (Li, 1997), ⑤ the Dahongshan Group (Hu et al., 1991; Greentree and Li, 2008), ⑥ the Baoban Complex and the Shilu Group (Li Z X et al., 2008a), ⑦ the Kunyang Group (Greentree et al., 2006), ⑧ the Huiqinggou granitic gneiss (Li Z X et al., 2002), ⑨ the Yanbian Group (Li et al., 2006b), though Sun W H et al. (2009) suggested a slightly younger age, ⑩ the NE Jiangxi ophiolitic complexes (Chen et al., 1991; Li et al., 1994; Li and Li, 2003; Li W X et al., 2008a), ⑪ the Shuangxiwu arc (Ye et al., 2007; Chen et al., 2009; Li et al., 2009), and its southwest extension, the Tianli Schists (Li et al., 2007), ⑫ the Xixiang arc (Ling et al., 2003)

1

accepted as a coherent Yangtze Block/Craton with some Archean basement. However, controversy still remains regarding the timing of collision between the Yangtze and the Cathaysia blocks, and the composition of the Cathaysia Block.

➢1.1 A brief reappraisal of tectonic models for southeastern SCB developed before the 1990s

The dominant tectonic model before the 1980s suggests that southeastern South China Block was an early Paleozoic fold belt developed over a miogeosyncline (now a disused term), recognizing the strong imprint of the Ordovician-Silurian orogenesis on thick volcanic-sedimentary successions in the region (Huang et al., 1980; Ren, 1991). The reason for proposing a miogeosyncline was that no early Paleozoic ophiolite complex had been documented in the region (this remains the case today).

An alternative model, first proposed by Grabau (1924) and attracted renewed interests since the 1980s (Shui, 1987), recognizes the existence of Precambrian basement in various parts of southeastern South China and the continental shelf, and calls the coastal region the Cathaysia paleocontinent, or more commonly known as the Cathaysia Block.

These two concepts are not mutually exclusive regarding the physical nature of the basement of the coastal region. These models evolved into a remnant ocean model (Shui, 1987), in which a remnant ocean existed between the Yangtze and Cathaysia blocks after their eastern ends first touched during the late Precambrian (Figure 1.2(a)). The progressive closure of this remnant ocean during the early Paleozoic caused the deformation and metamorphism in the region (Figure 1.2(b), (c)), commonly known in the Chinese literature as the "Caledonian" orogeny (see an updated view in Section 1.6). A similar model was adapted by Liu and Xü (1994).

Other, more radical models have also been developed since the late 1970s. These include arc/terrane accretion models which suggest that the southeastern half of the SCB was formed through successive accretion of arcs and continental terranes younging toward the coast, from the Mesoproterozoic to as young as the Mesozoic (Qiao and Geng, 1981; Guo et al., 1986; Li, 1992).

Hsü and his Chinese colleagues (Hsü et al., 1988; Hsü et al., 1990), recognizing the widespread Mesozoic thrusting in southeastern South China, proposed a two-plate Alps-style tectonic model involving the closure of a Mesozoic ocean between the Yangtze Block and a coastal terrane they called the Huanan Block (Figure 1.2(d), (e)). This model triggered one of the most heated debates on the tectonic history of south China. Although Hsü (1994) subsequently conceded that the key "Mesozoic Banxi mélange", upon which their two-plate model was based, is likely to be of late Precambrian age, he still insisted on the existence of a Mesozoic ocean-closure between the two plates.

The Shui (1987) model

(a) Earliest Cambrian   (b) End Cambrian   (c) Latest Ordovician to Silurian

The Hsü et al. (1988, 1990) model

(d) Paleozoic — Tr1   (e) Late Mesozoic

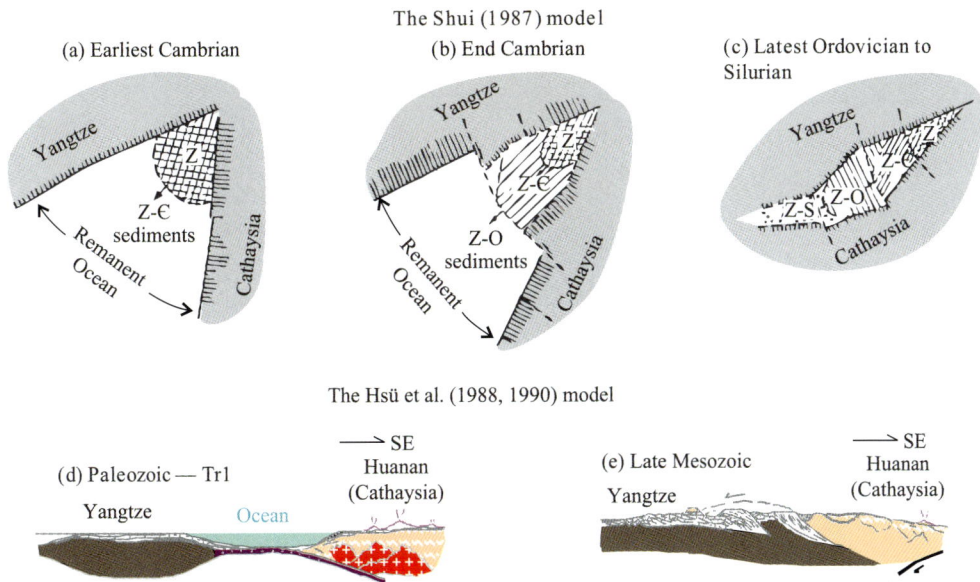

Figure 1.2   Schematic diagrams showing two early tectonic models for South China, modified from the original work by the proponents of each model

A common issue for many of the proposed models is that no reliable geological record exists for active margins in the continental interior since at least the end of the Precambrian. In addition, the distribution of sedimentary facies in South China (Liu and Xü, 1994) exhibits coherent patterns across the proposed Mesozoic sutures since at least the Devonian (Li, 1998).

A resurgence of investigations since the 1990s has seen a rapid accumulation of more systematic and reliable geochronology, geochemistry, and basin analysis results, and revision of the tectonic models in regional or global contexts. Some of these models will be explored/documented in the following sections.

➢1.2   A cartoon time-space diagram summarising major events that shaped the SCB

Figure 1.3 illustrates that distinct pre-Neoproterozoic tectonic histories were recorded by the Yangtze and Cathaysia blocks, and the two blocks started to share a common tectonomagmatic and basin record from the Neoproterozoic.

Major events include :

(1) Late Mesoproterozoic to early Neoproterozoic Sibao/Jiangnan Orogeny (1300–1200 Ma to 880 Ma) leading to the formation of the united SCB;

(2) Mid-Neoproterozoic (850–720 Ma) continental rifting and magmatism (Nanhua, Kangdian and Bikou-Hannan rift basin development), with a failed continental rift basin (the Nanhua Basin) remaining at the end of the Neoproterozoic;

3

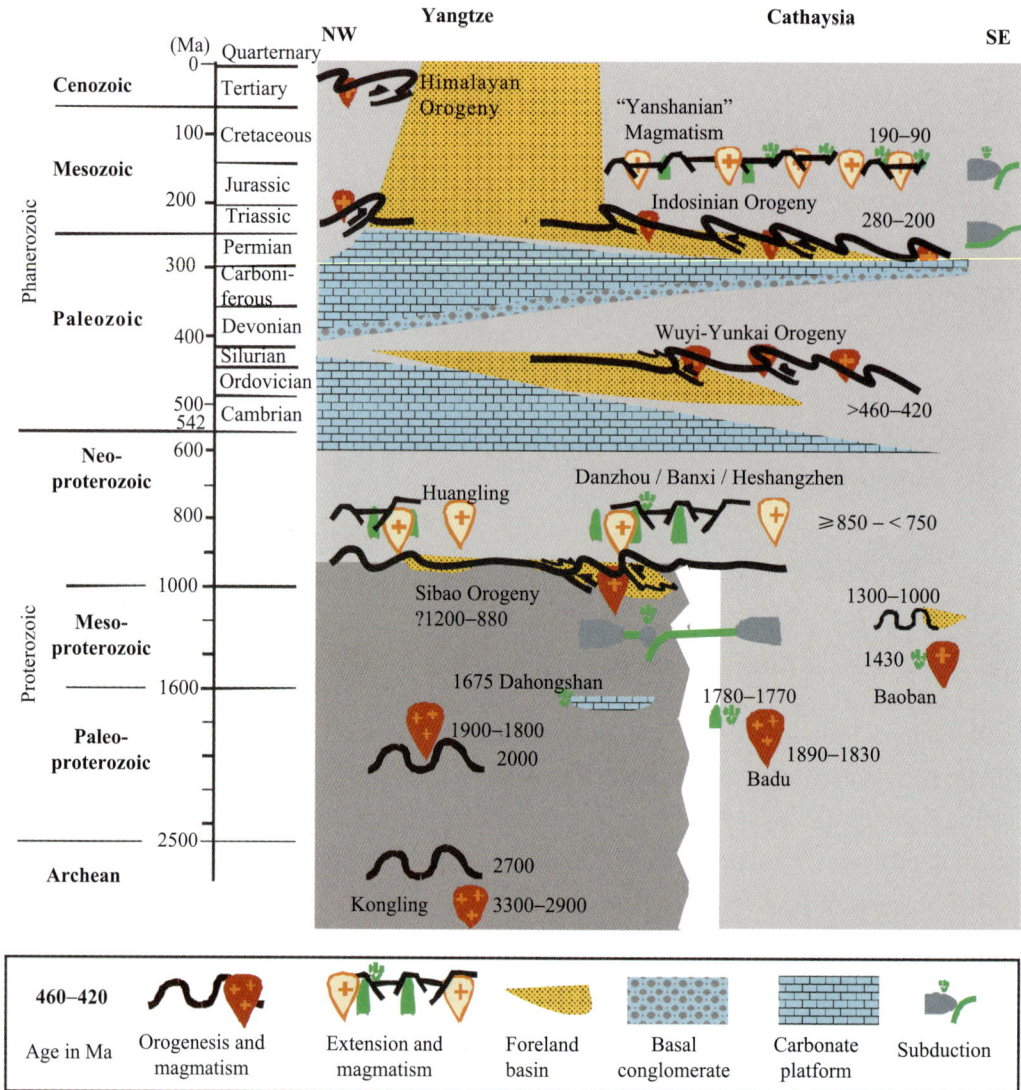

Figure 1.3   A schematic diagram illustrating the timing of major tectonic events in South China

(3) Early Paleozoic Wuyi-Yunkai Orogeny that started from the Cathaysia Block and advanced toward the Yangtze Block, accompanied by widespread, mostly late orogenic magmatism;

(4) Late Paleozoic development of platform cover sequence across the SCB, featuring a marine transgression started from the western side of the continent;

(5) Permo-Triassic Indosinian Orogeny, syn- to late-orogenic sag basin development, and post-orogenic ("Yanshanian") extension and magmatism;

(6) Development of the Sichuan Basin as a three-sided foreland basin since the Mesozoic;

(7) Impact of the Himalaya Orogeny and the opening of the South China Sea in the Cenozoic.

Multiple models exist to explain the cause of most of these events. Even the age ranges of some of the events are controversial. In the following sections, we will give our preferred account of some of the major events based on both field observations and analytical results, and will only briefly touch on some of the alternative interpretations. Please refer to original papers for more complete discussions of the various models.

➤1.3   Pre-1 Ga basement compositions

Outcrops of pre-Neoproterozoic crystalline basement are scarce in South China, with the oldest being the Kongling Complex in the Yangtze Block ("1" in Figure 1.1), which has ca. 3.3–3.2 Ga and 2.95–2.90 Ga igneous (including trondhjemitic) rocks that experienced ca. 2.75 Ga high-grade metamorphism, and 1.9–1.8 Ga granitic intrusions (Qiu et al., 2000; Zhang et al., 2006; Jiao et al., 2009; Gao et al., 2011). Paleoproterozoic rocks were thought to exist along the western margin of the Yangtze Block (e.g., the Kangding Complex), but recent dating indicate that they are either metamorphosed ca. 1000 Ma sediments (Li Z X et al., 2002), or ca. 800–745 Ma intrusions (Zhou et al., 2002b; Li Z X et al., 2003a). The 1.68 Ga Dahongshan Group metavolcanic-sedimentary succession is found in the southwestern corner of the Yangtze Block only ("5" in Figure 1.1; Greentree and Li, 2008; Hu et al., 1991), and more 1.7–1.5 Ga rocks have recently been reported in nearby regions (Sun W H et al., 2009; Zhao et al., 2010; Fan et al., 2013). Sedimentary protolith of the Tianli Schists, with an deposition age no older than its youngest detrital zircon population of 1530 Ma, is also thought to have formed on the southern margin of the Yangtze Block during late Mesoproterozoic ("11" in Figure 1.1; Li Z X et al., 2007).

There have been documented differences in the isotopic signatures between the Yangtze and the Cathaysia blocks (e.g., Chen and Jahn, 1998, Figure 1.4). However, the exact boundary between the two, particularly over the western half of the SCB, remains unclear. In eastern SCB, the early Neoproterozoic Shuangxiwu arc and its extension to the Tianli Schists ("11" and "11'" in Figure 1.1 Li Z X et al., 2007; Ye et al., 2007; Li et al., 2009b) likely mark the southern margin of the Yangtze Block.

Although Archean detrital zircon grains have been reported from Cathaysian rocks in numerous studies (Li, 1997; Wan et al., 2007; Xu et al., 2007; Yu et al., 2009; Li Z X et al., 2010), no Archean rock has been identified so far. The oldest known crystalline rocks are the ca. 1.9–1.8 Ga granitic rocks and basalts (metamorphosed to amphibolite facies) in western Zhejiang ("2" and "3" in Figure 1.1) and northwestern Fujian ("4" in Figure 1.1). On the Hainan Island, granites and metavolcaniclastic rocks are dated at ca. 1.43 Ga ("6" in Figure 1.1; Ma et al., 1998; Li Z X et al., 2002; Li Z X et al., 2008a). There are other widespread, mostly high-grade metamorphic rocks in northeastern Cathaysia (regions between "3" and "4" in Figure 1.1) that were thought to be part of the Precambrian crystalline basement of Cathaysia, but recent SHRIMP work indicate that they are Neoproterozoic rift successions metamorphosed during the Phanerozoic (Wan et al., 2007; Li Z X et al., 2010).

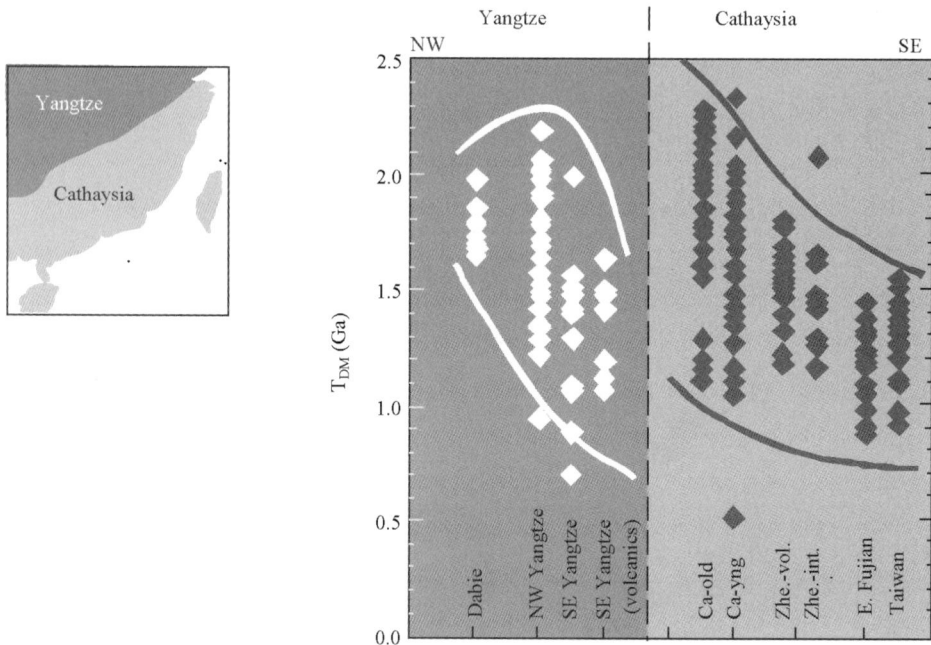

Figure 1.4   Nd model age (T_{DM}) of igneous rocks in South China (after Chen and Jahn, 1998)
Ca – Cathaysia; Zhe. – Zhejiang

➢1.4   Mesoproterozoic to earliest Neoproterozoic Sibao Orogeny: Part of Rodinia assembly?

Most researchers now accept that the Yangtze and Cathaysia blocks joined together by Late Neoproterozoic time, but models differ regarding the timing of this amalgamation and how it occurred. One group suggests that the amalgamation occurred diachronously, first at the SW-ends of the two continental blocks at ca. 1140 Ma (Greentree et al., 2006; Li Z X et al., 2002), and eventually at the NE-ends of the two blocks by 900–880 Ma (Li Z X et al., 2007; Li W X et al., 2008a; Li et al., 2009). In the western Sibao Orogen (also known as the Jiangnan Orogen), the first evidence of sedimentary provenance linkage between the two blocks is the appearance of detrital zircons of possible Cathaysia origin in part of the traditionally defined Kunyang Group (the Laowushan Formation, deposition age 1142 ± 16 Ma, "7" in Figure 1.1; Greentree et al., 2006). Siliciclastic rocks with likely Cathaysia contributions were also found in ca. 1000 Ma sediments in southern Sichuan ("8" in Figure 1.1), where a 1007 ± 14 Ma granitic gneiss was also found (Li Z X et al., 2002). A post-orogenic granodiorite dated at 857 ± 13 Ma (Li X H et al., 2003b) provides the younger age limit for the orogenic events there. On the Cathaysia side, ca. 1300–1000 Ma amphibolite-facies metamorphism occurred in the Hainan Island ("6" in Figure 1.1; Li Z X et al., 2002), but whether the region was part of the Sibao Orogen is still unclear.

In eastern Sibao Orogen, plutonic and volcanic rocks in the Shuangxiwu arc ("11" in Figure 1.1) have been dated at between ca. 970 Ma and 890 Ma (Chen et al., 2009; Li et al., 2009; Ye et al., 2007). The deformed arc was intruded by 849 ± 7 Ma post-orogenic Shenwu dolerite dykes (Li X H et al., 2008), and unconformably overlain by the ca. 800 Ma Luojiamen Group that is interpreted to represent Neoproterozoic rifting (Wang and Li, 2003). The Tianli Schists, a possible SW-extension of the Shangxiwu arc ("11" in Figure 1.1), recorded metamorphism and structural reactivations between ca. 1040 Ma and 940 Ma (Li Z X et al., 2007), and is unconformably overlain by 827 ± 14 Ma bimodal rift magmatism (Li W X et al., 2008b). The NE Jiangxi ophiolitic complexes ("10" in Figure 1.1; Zhou, 1989) started to form at ca. 1000–970 Ma (Chen et al., 1991; Li et al., 1994; Li and Li, 2003) and were obducted at ca. 880 Ma (Li W X et al., 2008a). Sibaoan arc complexes (≥900 Ma) also exist along the northern margin of the Yangtze Block ("12" in Figure 1.1; Ling et al., 2003).

Meso- to Neoproterozoic orogenic events in South China thus appear to have an age range of 1300–880 Ma, with much of the suturing between the Yangtze and Cathaysia blocks occurred after ca. 1000 Ma (Figure 1.5). Worldwide orogenic events of this time interval, which are often referred to as "Grenvillian" orogenic events but in places lasted longer than the classic Grenville Orogeny in Laurentia (Davidson, 1995; Rivers, 1997), are believed to be responsible for the assembly of the Neoproterozoic supercontinent Rodinia (Dalziel, 1991; Hoffman, 1991; Moores, 1991; Li Z X et al., 2008b).

Recognising the similarity of the Cathaysia crustal provinces with that of southern Laurentia, and the requirement for a western source region similar to Cathaysia for upper Belt Basin deposits (Ross et al., 1992), Z. X. Li and co-workers (Li Z X et al., 1995b, 2002, 2008a, 2008b) suggested that the Cathaysia Block was part of southwestern Laurentia (Figure 1.6), and that the Sibao Orogeny led to the suturing between Laurentia-Cathaysia and the Yangtze Block (Figure 1.5) during the final assembly of Rodinia (Figure 1.7(a), (b)). Such a configuration is consistent with comparable Neoproterozoic rift histories and plume activities between Australia, South China and western Laurentia (see next section). It also provides an answer for geological mismatches in the classic SWEAT configuration for Rodinia (Borg and DePaolo, 1994), because it has Sibaoan sutures between Australia-East Antarctica and Laurentia. The reconstructions as in Figure 1.7(a), (b) are consistent with currently available paleomagnetic results (see reviews in Li Z X et al., 2008b).

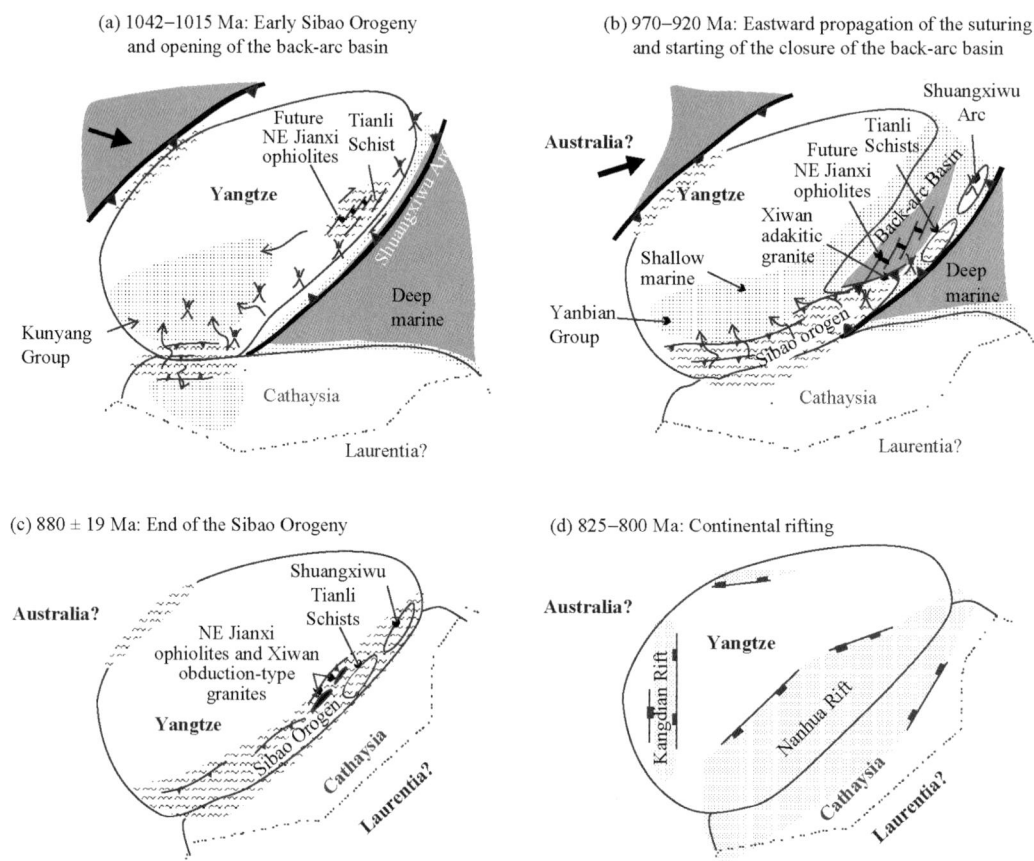

Figure 1.5 A schematic diagram illustrating a possible process of amalgamation between the Yangtze and Cathaysia blocks and the change of tectonic regime into an extensional environment by ca. 850 Ma (Li Z X et al., 2007; Li et al., 2009)

There are also other current tectonic models that imply a ca. 820–800 Ma age for the amalgamation between the Yangtze and Cathaysia blocks (Li, 1999; Zhao and Cawood, 1999; Wang et al., 2006; Wu et al., 2006) or even younger (Zhou et al., 2002b). However, all these models suffer from the lack of post-890 Ma typical arc magmatism, the lack of post-880 Ma metamorphic event related to such a convergent event, and the wide occurrence of ≤850 Ma continental rifting and plume magmatism (see next section).

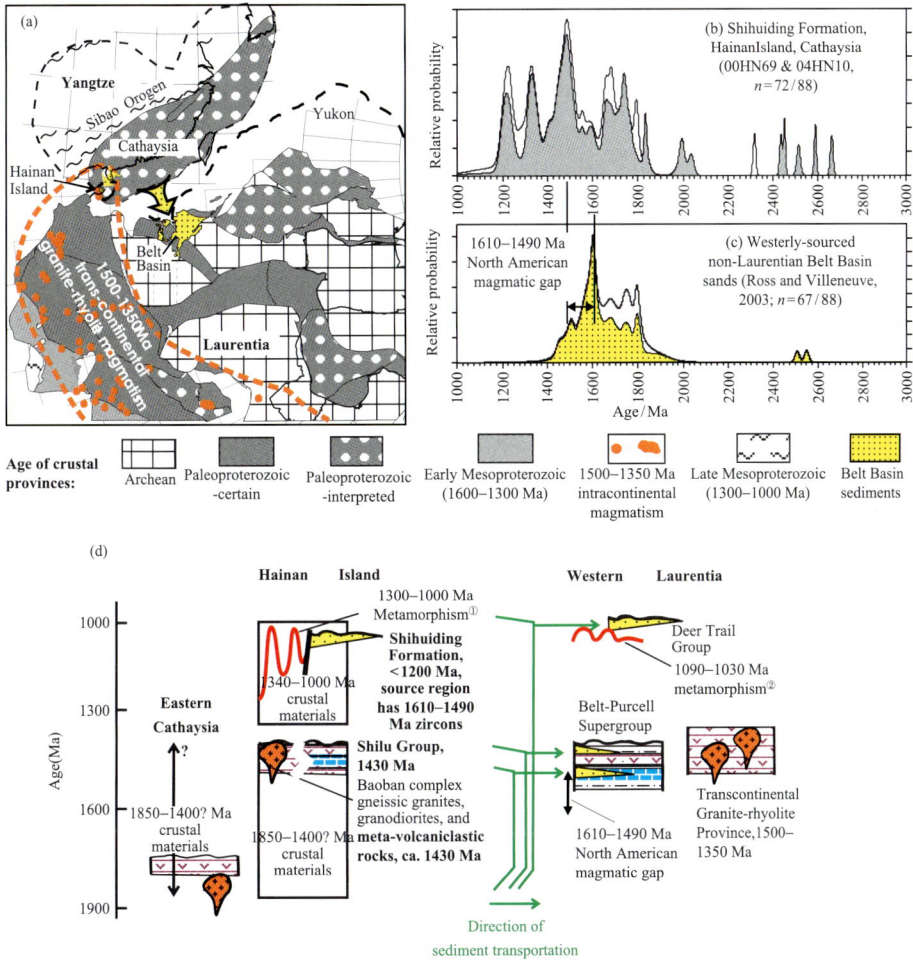

Figure 1.6　(a) Correlations of crustal provinces between Laurentia and Cathaysia (Li Z X et al., 2008a). Rotated present-day coordinates are shown in 5° intervals (thin doted lines); (b), (c) show the correlation of detrital provenance data between Hainan Island of Cathaysia and the westerly-sourced non-Laurentian sands in the Belt Basin of western Laurentia (south-west North America); (d) A schematic diagram illustrating sedimentary provenance linkages and tectonostratigraphic correlations between Mesoproterozoic Hainan Island and south-western Laurentia. The Cathaysia Block is interpreted as a source of sediments for westerly-derived sand wedges in western Laurentia

Data sources:　① Li Z X et al. (2002);　② Anderson and Davis (1995)

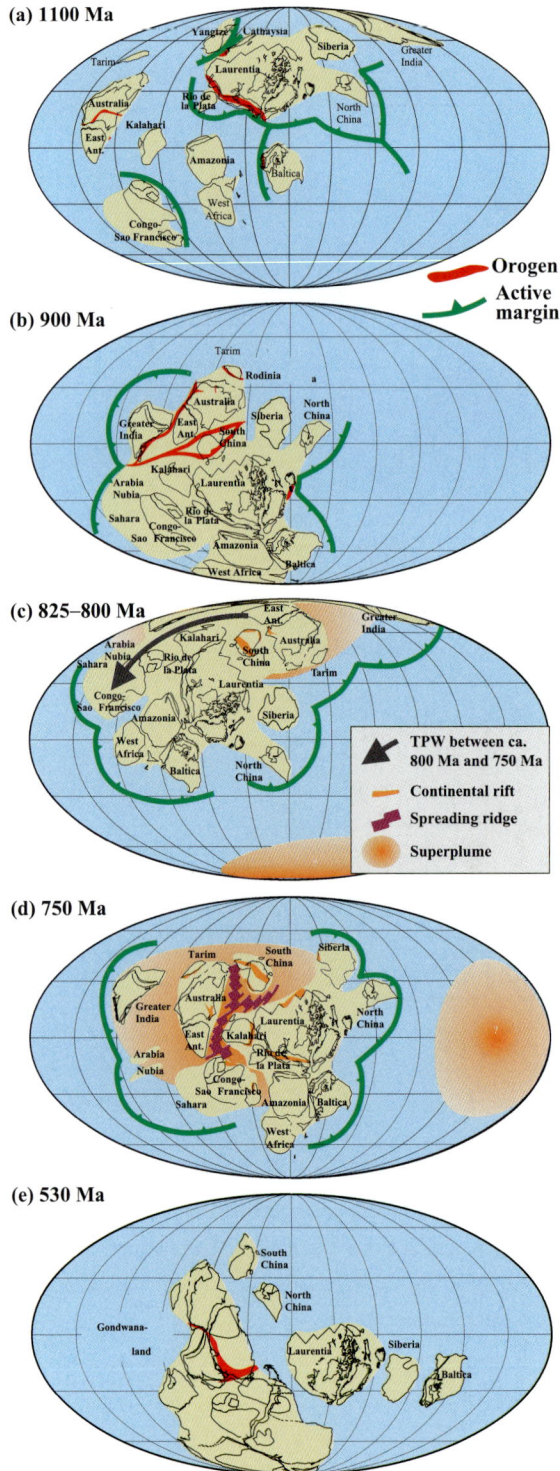

Figure 1.7 The formation and breakup of Rodinia during late Precambrian, and the formation of Gondwanaland by the Early Cambrian (Li Z X et al., 2008b)

➤1.5   Large scale mid-Neoproterozoic intraplate magmatism and continental rifting: Long-lived mantle plume activities and Rodinia breakup[*]

Neoproterozoic magmatism, including plutons, dyke swarms and volcanics, are widespread in South China (Figure 1.8). However, their tectonic significance remains controversial. The traditional view was that the plutons signify the "cratonization" of the Yangtze Block (Wang and Mo, 1995). However, precise geochronology has demonstrated that a vast majority of the plutons are coeval with continental rifting events, predominantly between 825 Ma and 750 Ma (Liu, 1991; Li et al., 1999; Li X H et al., 2003a; Li Z X et al., 2003a; Wang and Li, 2003).

Li Z X et al. (2003a) demonstrated that after a possible early start at ca. 860–850 Ma, Neoproterozoic magmatic events in South China exhibit four major peaks: ca. 825 Ma, ca. 800 Ma, ca. 780 Ma and ca. 750 Ma. More ages on all age groups have been reported in recent years. In particular, Li X H et al. (2003b; 2006b, 2008, 2010), Li W X et al. (2010), Li Z X et al. (2010) and Shu et al. (2011) demonstrated that anorogenic magmatism, including ca. 850 Ma bimodal rift magmatism (Li W X et al., 2010; "16" in Figure 1.8), started at 860–850 Ma in South China. Although the anorogenic interpretation for the 860–750 Ma magmatism is not universally accepted (for alternative views see Wang et al., 2006 and Zhou et al., 2002c), there are strong evidences indicating that these rocks are products of plume-induced magmatism, particularly those dated at 825–750 Ma (see discussions below), 860–850 Ma magmatism (Li et al. 2006b; Li Z X et al., 2008b; Li X H et al., 2010; Shu et al., 2011) are not as widespread or well studied. Some of the magmatic events are also found in the southern Korean Peninsula, an extension of the South China Block.

1.5.1   Ca. 825 Ma magmatism: the Guibei large igneous province (LIP)

The Guibei LIP covers products of a major bimodal magmatic event in South China. Mafic-ultramafic rocks include the 828 ± 7 Ma mafic-ultramafic dykes in Guibei (marked as "1" in Figure 1.8; Li et al., 1999), the ca. 825–810 Ma Tongde-Gaojiacun Complex ("2" in Figure 1.8; Sinclair, 2001; Zhou et al., 2006; Zhu et al., 2006), the 821 ± 7 Ma to 811 ± 12 Ma Bikou Group basalts ("3" in Figure 1.8; Wang et al., 2008), the 821 ± 7 Ma Tiechuanshan basalts ("4" in Figure 1.8; Ling et al., 2003), the 820–810 Ma Wangjiangshan Complex ("5" in Figure 1.8; Zhou et al., 2002a), the 826 ± 3 Ma Yiyang komatiitic basalts ("6" in Figure 1.8; Wang et al., 2007, 2009), the 827 ± 4 Ma basalts at Guangfeng ("7" in Figure 1.8; Li W X et al., 2008b), and the 818 ± 9 Ma Mamianshan basalts ("8" in Figure 1.8; Li et al., 2005). There are also numerous granitic intrusions of that age in both the interior and along the margins of the Yangtze Block, and felsic volcanic and volcani-clastic rocks in the continental rift systems (Li

---

[*] modified from Li Z X et al., May 2009 LIP of the Month: http://www.largeigneousprovinces.org/09may

X H et al., 2003a).

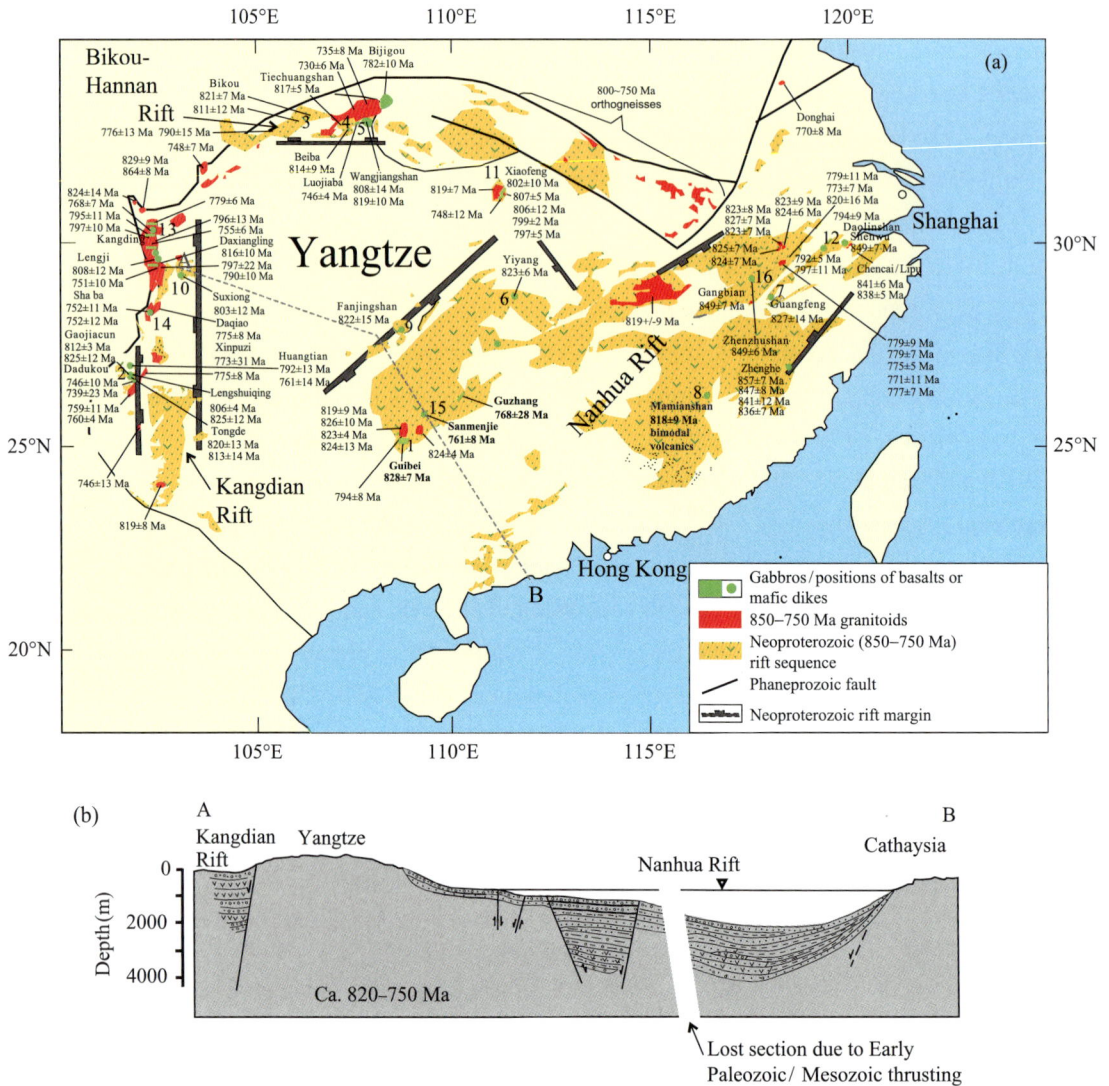

Figure 1.8 (a) Distribution and ages of 860–750 Ma magmatic rocks and continental rift systems in South China (Li X H et al., 2003a; Li et al., 1999; Li Z X et al., 2003a; Wang et al., 2010); (b) A schematic diagram showing the Kangdian and Nanhua rift systems (Liu and Xü, 1994)

The largest basaltic outcrop is the Bikou Group volcanics close to the present northwestern margin of the South China-Block ("3" in Figure 1.8). It covers an area of about 10,000 km^2, with an estimated kilometres-thick of predominantly tholeiitic basalts (Wang et al., 2008). Well-preserved pillow structures are found in several successions (Figure 1.9).

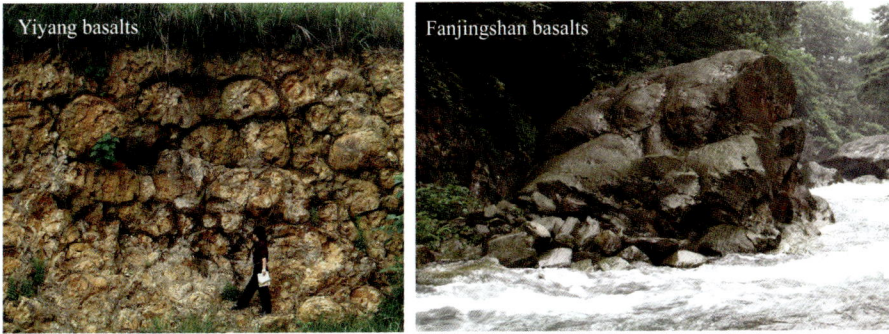

Figure 1.9   Pillow structures in the Yiyang komatiitic basalts ("6" in Figure 1.8; Wang et al., 2007) and the Fanjingshan basalts ("9" in Figure 1.8) (Photographs in this and following figures were all taken by Li Z X)

A mantle plume model for this episode of bimodal magmatism was based on several geological, geochemical and geochronological observations, such as evidence for syn-magmatic doming, the bimodal and intraplate nature of the magmatism (see summary geochemical characteristics below), and their associations with continental rifting (Li X H et al., 2003a; Li et al., 1999; Li Z X et al., 2003a). More recent work by Wang et al. (2007, 2008) demonstrate that both the 826 ± 3 Ma Yiyang **komatiitic basalts** and the 811 ± 12 Ma upper Bikou basalts have **mantle potential temperatures** over 1500°C (Figure 1.10).

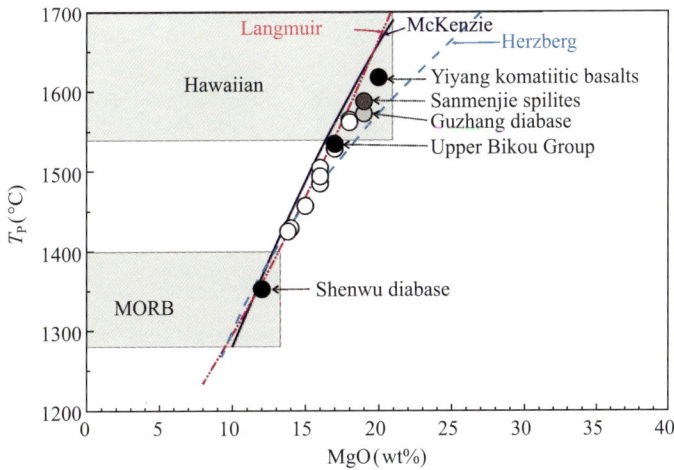

Figure 1.10   Mantle potential temperatures (T_P) as a function of the MgO concentrations of primary magmas (Wang et al., 2009, modified after Herzberg et al., 2007)

Zhou et al. (2009) recently reported 822 ± 15 Ma basalts at Fanjingshan ("9" in Figure 1.8; Figure 1.9), but they interpreted the rocks to be of arc origin. However, we notice that the basalts show no Zr and Hf depletion relative to Sm (as in Figure 8(c), (d) of Zhou et al., 2009), unlike typical arc basalts (McCulloch and Gamble, 1991). The Nb-Ta negative anomalies in the Fanjingshan basalts are thus likely results of crustal or sub-continental lithospheric mantle contamination (Li X H et al., 2007c; Wang et al., 2009). Indeed, plume-related low-Ti continental flood basalts (CFBs) often show Nb-Ta negative anomalies (Turner and Hawkesworth, 1995; Hawkesworth et al., 2000; Puffer, 2001). The Fanjingshan basalts are also alkaline basalts, and, therefore, should not be classified as calc-alkiline basalts.

1.5.2   Ca. 800 Ma magmatism: the Suxiong-Xiaofeng LIP

The Suxiong-Xiaofeng LIP is represented by the 803 ± 12 Ma (Figure 1.11) Suxiong-Kaijianqiao bimodal volcanics in the Kangdian Rift ("10" in Figure 1.8; Li X H et al., 2002; Wang and Li, 2003), the 802 ± 10 Ma Xiaofeng composite dykes ("11" in Figure 1.8; Li X H et al., 2004), the 794 ± 9 Daolinshan granite-diabase complex and the 792 ± 5 Ma Shangshu bimodal (basalt-rhyolite) volcanic rocks ("12" in Figure 1.8; Li X H et al., 2008), and equivalent plutonic and volcanic rocks elsewhere. The Suxiong-Kaijianqiao bimodal volcanic succession is up to 10 km thick, and the Shangshu Group, dominately basalts, is up to 2.5 km thick.

Figure 1.11   The 802 ± 10 Ma Xiaofeng dykes intruding the 819 ± 7 Ma Huangling granite (Li et al., 2004)

1.5.3   Ca. 780 Ma magmatism: the Kangding LIP

Products of the Kangding events including the roughly E-W trending, 779 ± 6 Ma to 768 ± 7 Ma mafic dykes that intrude coeval granitic rocks in the Kangding region ("13" in Figure 1.8, also Figure 1.12; Li Z X et al., 2003a; Lin et al., 2007), and numerous coeval granitic and volcanic units throughout the South China Block (Figure 1.8).

1.5.4   Ca. 750 Ma magmatism: the Shaba LIP

The ca. 750 Ma event was widespread in South China, with mafic examples include the Shaba gabbro ("14" in Figure 1.8; Li Z X et al., 2003a; Lin et al., 2007), the Sanmenjie spilites and gabbros, and the Guzhang mafic dykes ("15" in Figure 1.8; Zhou et al., 2007). Both the Sanmenjie spilites and the Guzhang mafic dykes indicate high mantle potential temperatures (Figure 1.10; Wang et al., 2009).

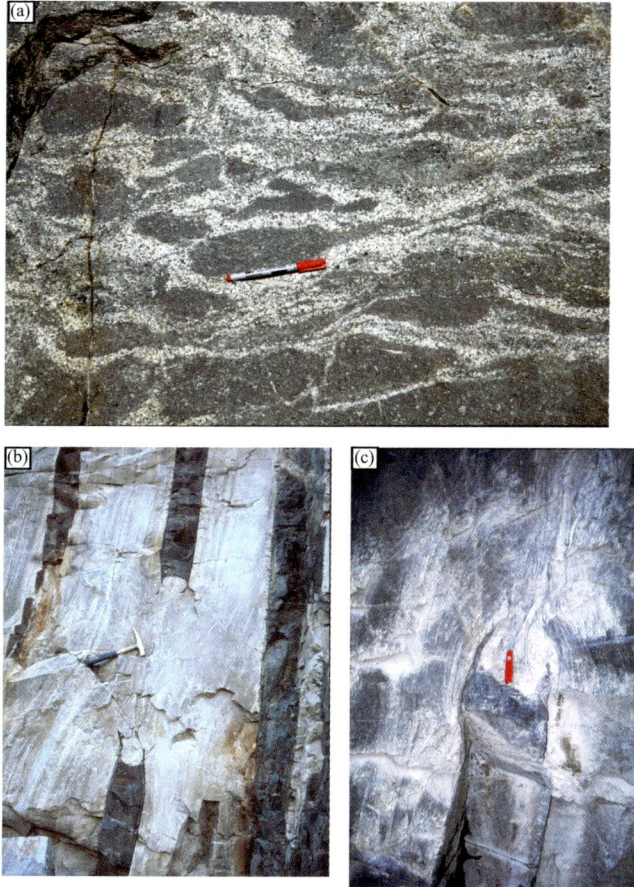

Figure 1.12   Photos of 779 ± 6 Ma to 768 ± 7 Ma magma mingling (a), and ductile interaction between mafic dykes and the intruded granitoid when they both were in semi-solid states (b, c) in the Kangding region (Li Z X et al., 2003a; Lin et al., 2007)

1.5.5   Geochemical characteristics of the 860–750 Ma magmatic rocks in South China

The 860~750 Ma basaltic rocks are predominantly sub-alkaline series and subordinately alkaline series (Figure 1.13(a)). The sub-alkaline basalts are dominantly of tholeiitic nature, and calc-alkaline basalts are rare (Figure 1.13(b)). Fe-poor, Si-enriched and Nb-Ta-depleted characteristics of some basaltic rocks reflect contributions from SCLM. Nonetheless, the anhydrous high-temperature lavas and typical OIB-type basalts are believed to have sampled a plume mantle source.

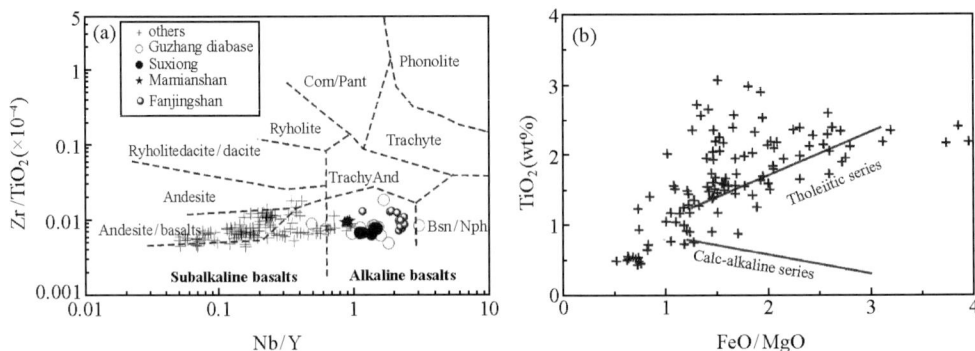

Figure 1.13 (a) Nb/Y vs. $Zr/TiO_2 \times 10^{-4}$ diagram distinguishing sub-alkaline and alkaline basalts; (b) FeO/MgO vs. SiO_2 diagram distinguishing tholeiitic and calc-alkaline series (Wang et al., 2009)

1.5.6　Geodynamic model

Repeated mantle plume events appear to best explain the geological and geochemical characteristics of the 825–750 Ma igneous rocks in South China, their wide distribution, and association with continental rifting (Figure 1.14).

Similar Neoproterozoic records are found across a number of continents during Rodinia time (e.g., Ernst et al., 2008; Li Z X et al., 2008b; Li Z X et al., 2003a). Li Z X et al. (2003a; 2008b) proposed that a mantle superplume beneath Rodinia was responsible for the widespread, and long-lasting (ca. 100 Ma) magmatic events over Rodinia that eventually caused the breakup of the supercontinent (Figure. 1.7(c), (d); Figure 1.15).

➢1.6　The Ordovicion–Silurian Wuyi-Yunkai ("Caledonian") Orogeny and foreland basin development

Like in Australia, continental rifting in South China ceased by ca. 700 Ma. Whereas the Neoproterozoic Kangdian Rift was filled up and platform carbonate deposition started following the Nantuo glaciation, the Nanhua basin continued to receive sediments after that (Liu and Xü, 1994; Wang, 1985). It is unclear whether any oceanic crust was developed in the Nanhua basin, but so far no post-Neoproterozoic arc or ophiolitic complexe have been documented between the Yangtze and Cathaysia blocks. It is interesting to note that sediments of Neoproterozoic Ⅲ to early Ordovician ages are dominantly carbonates, shales and some cherty units on the Yangtze side of the Nanhua basin, whereas clastic marine deposits dominate the Cathaysia side of the basin.

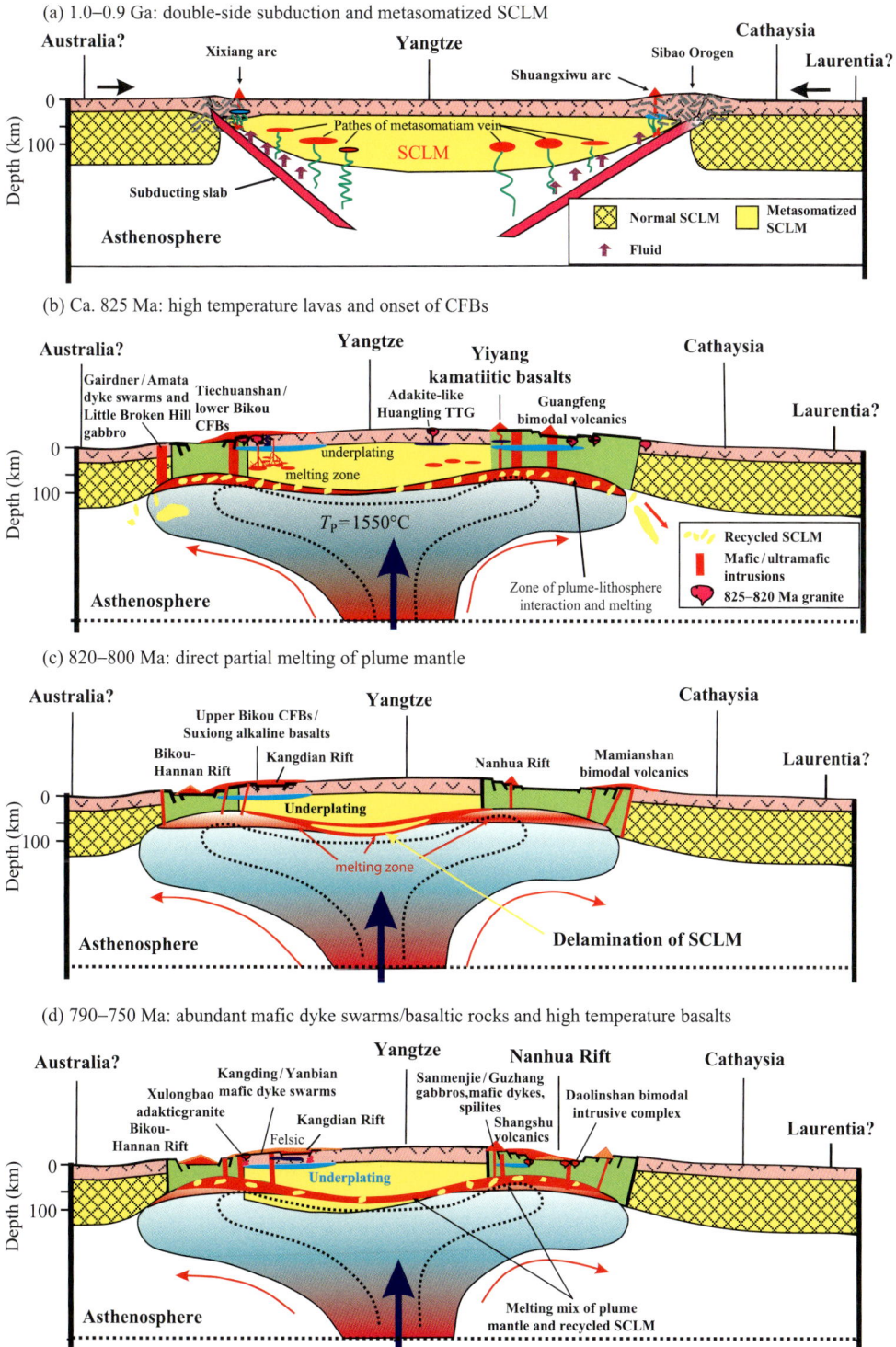

Figure 1.14 A geodynamic model for early Neoproterozoic South China (modified after Li et al., 1999; Wang et al., 2009)

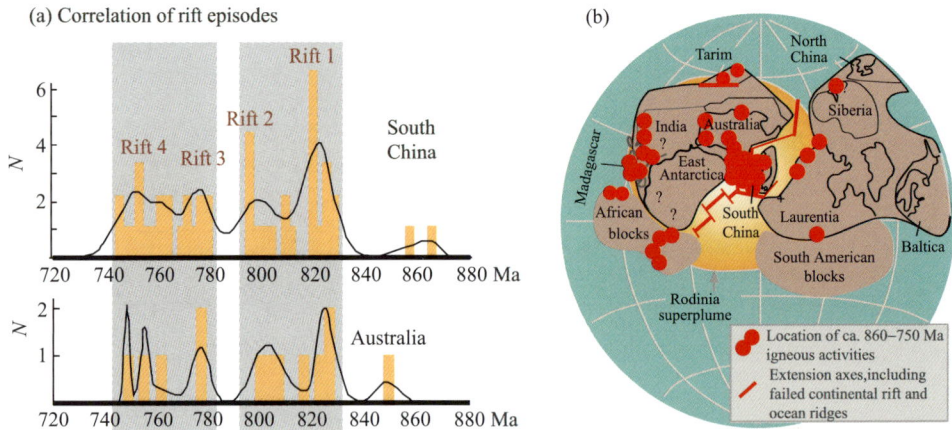

Figure 1.15 (a) Well correlated Neoproterozoic rifting events between South China and Australia, and (b) the Rodinia superplume model (after Li Z X et al., 2003a). N = number of precise ages available before 2003. Note that the 2003 Rodinia reconstruction in (b) is slightly different from that in Figure 1.7 published in 2008

An Early Paleozoic orogenic belt, traditionally called the "Caledonian Orogen" in Chinese literature but was termed the Wuyi-Yunkai Orogen by Li Z X et al. (2010), was developed in South China (Figure 1.16). The orogen covers the southeastern half of the South China Block, stretching for ca. 2000 km in a northeasterly direction (Ren et al., 1997) and is up to 600 km in width (Figure 1.16). It could have extended as far as the Korean peninsula (Charvet et al., 1999; Kim et al., 2006) and to the Indochina Block (Ren, 1991; Carter et al., 2001; Nagy et al., 2001; Roger et al., 2007). The orogen has been poorly studied in terms of its metamorphic history, magmatism, and its structural kinematics since its recognition half a century ago (Huang, 1960; Ren, 1964, 1991; Huang et al., 1980; DOGNU, 1981; Jahn et al., 1990;).

The Wuyi-Yunkai Orogeny was originally identified to be of Early Paleozoic (Ordovician?–Silurian) age, largely due to the recognition of an angular unconformity between post-Silurian cover and strongly-deformed pre-Devonian strata with intruded granites in the southeastern half of South China (Ren, 1964, 1991; Huang et al., 1980) (Figure 1.17(a)). Widespread metamorphic rocks in the northeastern part of the orogen were considered by some earlier workers to be early Neoproterozoic or older, possibly related to the accretion of the Cathaysia and Yangtze blocks during the assembly of Rodinia (Shui et al., 1988; Zhao and Cawood, 1999). It is only recently that reliable early Paleozoic magmatic and metamorphic ages have been published, defining the orogeny to be between mid-Ordovician (>460 Ma) and the end of Silurian (420–415 Ma) (Li Z X et al., 2010; Figure 1.16).

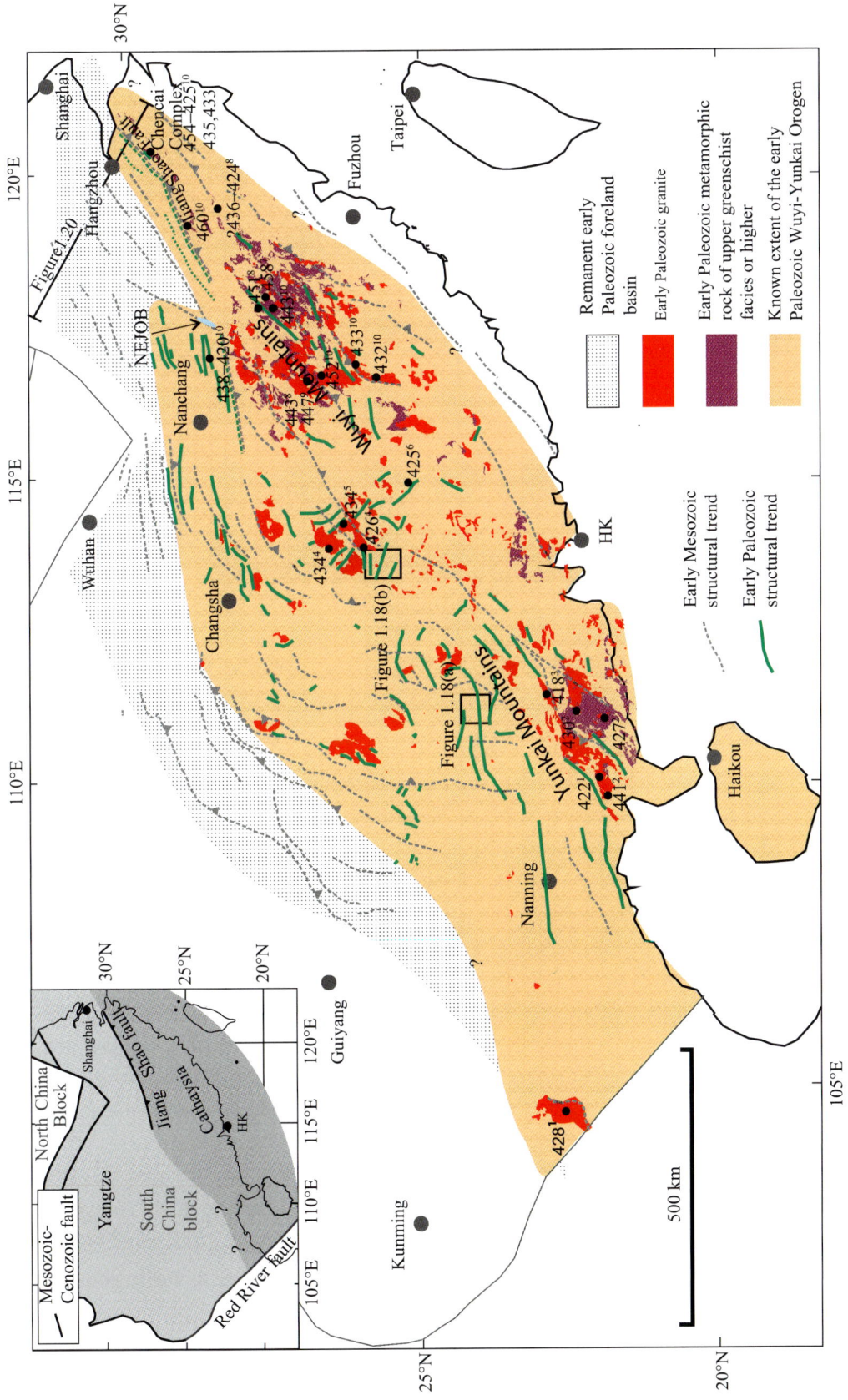

Figure 1.16 Simplified regional map highlighting the regional extent of the Early Paleozoic Wuyi-Yunkai Orogen, Early Paleozoic metamorphic and granitic rocks, and interpreted Early Paleozoic structural trends overprinted by Permian-Jurassic structures, with major crustal-scale thrusts shown (modified from Li Z X et al., 2010). Sources of listed ages (numbers next to black dots): 1–Roger et al. (2000); 2–Wang et al. (2007a); 3–Wang et al. (1998); 4–Li (1991); 5–Li et al. (1989); 6–Xu et al. (2005); 7–Wan et al. (2007); 8–Chen et al. (2008); 9–Zeng et al. (2008); 10–Li Z X et al. (2010). NEJOB – ca. 1 Ga NE Jianexi ophiolite belt (Zhou. 1989)

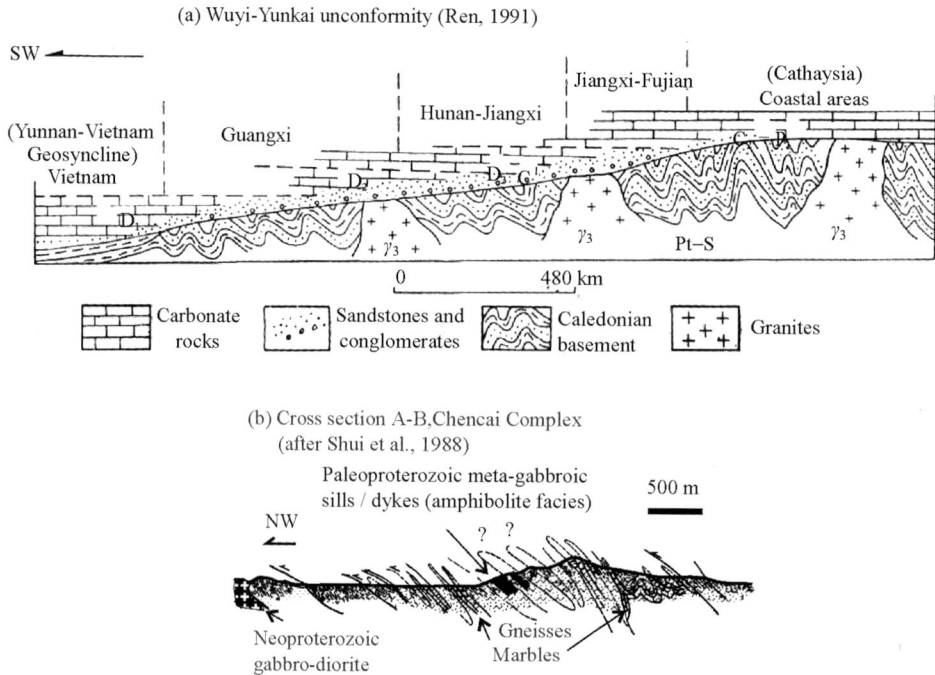

Figure 1.17 (a) Unconformity due to the Wuyi-Yunkai Orogeny; (b) Deformation style of the Chencai Complex (Figure 1.16 for location)

Early Paleozoic metamorphic rocks are commonly strongly deformed with predominantly northwestward vergence (Ren, 1964, 1991; see example of the deformation in the Chencai Complex in Figure 1.17(b)). This Early Paleozoic deformation also affected the cover rocks of the Yangtze Block in northwestern South China Block, but the intensity of deformation decreases toward the northwest (Ren, 1964) in a foreland fold-and-thrust-belt setting (Li, 1998; Li Z X et al., 2003b). Structures formed during the Wuyi-Yunkai Orogeny were overprinted by the Permian–Triassic orogenic event (e.g., Li and Li, 2007; Figure 1.16). In the northwestern half of the Wuyi-Yunkai Orogen, Early Paleozoic structural trends are commonly truncated by Permian–Triassic structures (Figure 1.16, Figure 1.18). However, over the southeastern half of the orogen, structural trends of the two orogenies are often sub-parallel, and are thus difficult to distinguish (Figure 1.16). Indeed, thrusts faults like the Jiang-Shao Fault were most likely active during both orogenies, although no systematic structural analysis and geochronology have yet been conducted to verify this.

In the eastern Wuyi-Yunkai Orogen, amphibolite-facies metamorphism occurred between >460 and 440 Ma, followed by a phase of rapid unroofing from >8 kbar[①] to ca. 4 kbar with a clockwise P-T path (Figure 1.19). The metamorphic rocks cooled to below 300–500°C by ca. 420 Ma (Li Z X et al., 2010). The clockwise metamorphic P-T path, similar to that established

---

① 1bar=10^5Pa

by Zhao and Cawood (1999), indicates crustal thickening during the initial stage of the orogen.

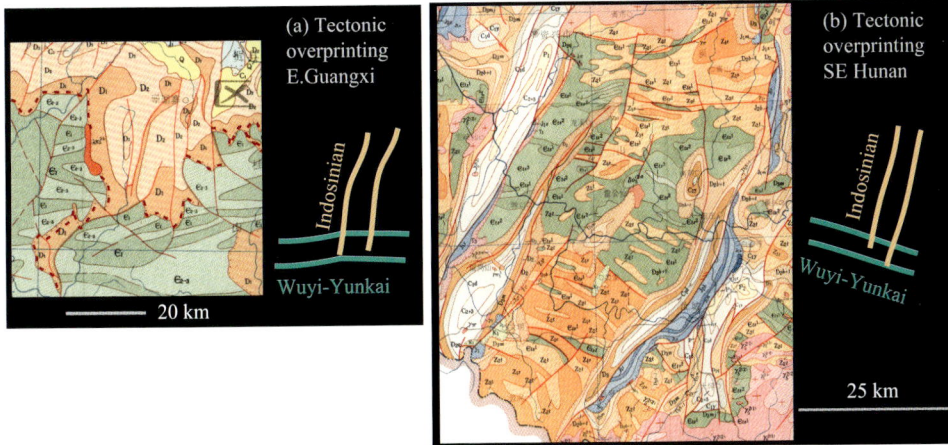

Figure 1.18   Overprinting of early Paleozoic (Wuyi-Yunkai) structures by Permian–Triassic (Indosinian) structures in central South China (after BOGAMR, 1984). Positions of the figures are shown in Figure 1.16

There has been no consensus regarding the tectonic environment and geodynamic driving force for the Wuyi-Yunkai Orogeny. The orogen was traditionally thought to be an Early Paleozoic miogeosyncline (Huang et al., 1980). Ren (1991) regarded the Wuyi-Yunkai Orogeny as an intraplate orogenic event that closed the scissor-shaped aulacogen, opening to the present-day south-west between the Yangtze and Cathaysia Blocks (Figure 1.2(a) – (c)). Li and co-workers (Li, 1998; Li Z X et al., 2003b) extended this model and regarded the orogeny to be of an intraplate nature, which closed the mid-Neoproterozoic–early Paleozoic Nanhua failed rift basin – the Nanhua Basin. Li (1998) further suggested that this intraplate orogeny was related to South China's interaction with the Indo-Australian margin of Gondwanaland. Similar models have been adapted by others, such as Wang et al. (2007a) and Charvet et al. (2010). The Nanhua Rift was most active during the mid-Neoproterozoic and was converted into a foreland basin (and its southern part become part of the Wuyi-Yunkai Orogeny) during the Middle Ordovician–Silurian, with deposition centers migrated toward the present-day northwest (Li, 1998; Figure 1.20). Such a model is compatible with: ① the well-documented Neoproterozoic rift events in southeastern South China (Li et al., 2005; Wang and Li, 2003); ② the lack of Early Paleozoic ophiolitic rocks or arc magmatism in the orogen; ③ the apparent exchange of sedimentary sources between the Yangtze and Cathaysia Blocks on the opposing side of the Nanhua basin during the Cambrian when the basin is supposed to be at its broadest (after rifting failed but prior to the onset of the Wuyi-Yunkai Orogeny; Li, 1998, after Liu and Xü, 1994); and ④ geochemical signatures of Phanerozoic granites (including the Early Paleozoic granites) indicating that most of these granites are anatectic products of Precambrian continental crust with little evidence for input from a subducting oceanic plate (Chen and Jahn, 1998; Jahn et al., 1990; Li and Gui, 1991). However, this does not rule out the possibility of a narrow strip of oceanic crust, like the present Red Sea, being present in the Nanhua rift system during the Neoproterozoic–Early Paleozoic. The latter scenario is similar to the model proposed by Liu and Xu (1994) in which the Wuyi-Yunkai Orogeny was due to the closure of a small oceanic basin between the Yangtze and Cathaysia blocks, with the oceanic plate subducting towards the Cathaysia Block.

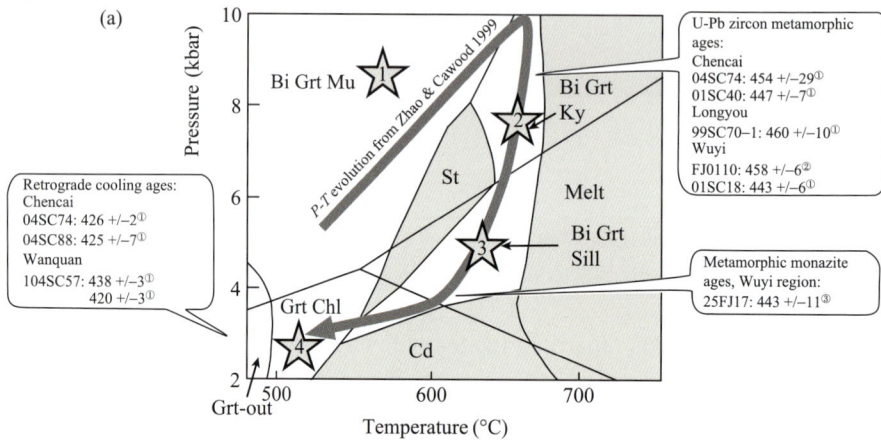

Figure 1.19 (a) *P-T-t* path of metamorphic rocks in eastern Wuyi-Yunkai Orogen, and (b) a typical Chencai paragneiss outcrop at Wuzili. Numbered stars show conditions estimated for the Chencai complex (Li Z X et al., 2010), and the *P-T* path was from Zhao and Cawood (1999) for metamorphic rocks in the Wuyi region. Sources of ages: ① Li Z X et al. (2010), ② Wan et al. (2007), ③ Chen et al. (2008)

Through combining geochronological data with preliminary geochemical and metamorphic analyses, Li Z X et al. (2010) proposed a tectonic model as in Figure 1.21, featuring an Ordovician initiation of an intracontinental orogeny developed on a failed continental rift, and a ca. 440–420 Ma orogenic collapse with widespread late orogenic magmatism, caused by orogenic root delamination (see new evidence in Yao et al., 2012).

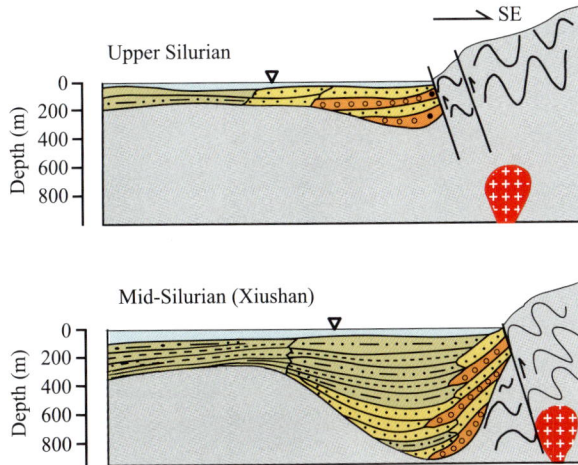

Figure 1.20 NW-ward propagation of the foreland basin toward the end of the Wuyi-Yunkai Orogeny (Li, 1998, after Liu and Xü, 1994). Position of the schematic cross-section is shown in Figure 1.16

➤1.7 Permian–Triassic orogenic events and a Jurassic–Cretaceous large magmatic province

A Devonian–Carboniferous marine transgression occurred in South China after the Wuyi-Yunkai Orogeny (Figure 1.17(a) and Figure 1.22). The entire SBC was covered by platform carbonate during Carboniferous–Early Permian time (Figure 1.22(c), (d)), leaving very little doubt that the South China Block had become a single continental block by then.

By the early to mid-Permian, continental collision between the South and North China blocks may have started at their eastern ends (Figure 1.23), as indicated by: ① the development of a fore-deep along the northern margin of the SCB (the Southern Qinling-Lower Yangtze Trough, Figure 1.24(b)), due to slab-pull and orogenic loading (Li, 1998); ② the first appearance of likely northerly-sourced clastic rocks in the Southern Qinling-Lower Yangtze Trough (Figure 1.24(a), (b)), indicating the possible development of a mountain belt in the north; ③ ca. 300–260 Ma metamorphic and cooling ages on both sides of the Sulu belt (Chen et al., 1992; Zhu et al., 1994; Wang et al., 1996); and ④ paleomagnetic results that permit such a collision (Zhao and Coe, 1987).

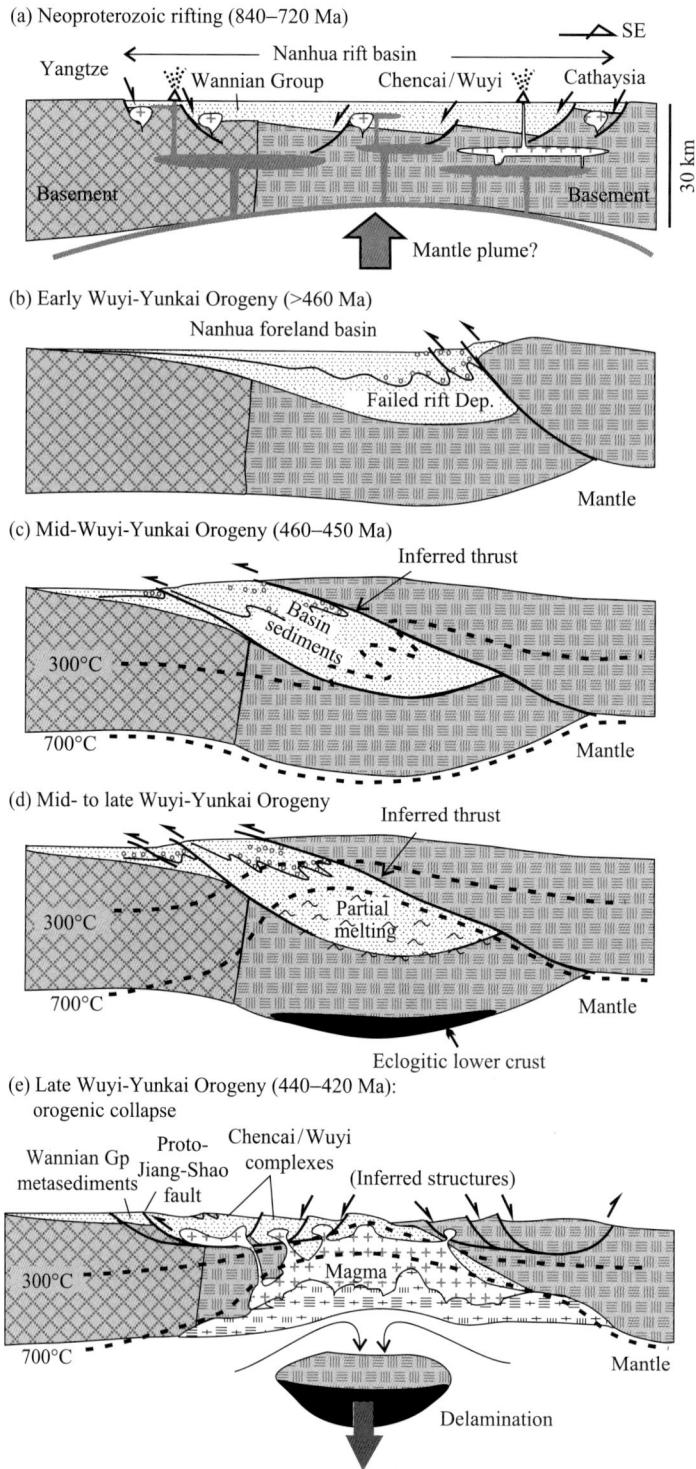

(a) Neoproterozoic rifting (840–720 Ma)

(b) Early Wuyi-Yunkai Orogeny (>460 Ma)

(c) Mid-Wuyi-Yunkai Orogeny (460–450 Ma)

(d) Mid- to late Wuyi-Yunkai Orogeny

(e) Late Wuyi-Yunkai Orogeny (440–420 Ma): orogenic collapse

Figure 1.21　A tectonic model illustrating the development of the Early Paleozoic Wuyi-Yunkai Orogen on the Neoproterozoic Nanhua failed rift basin (Li Z X et al., 2010; Yao et al., 2012)

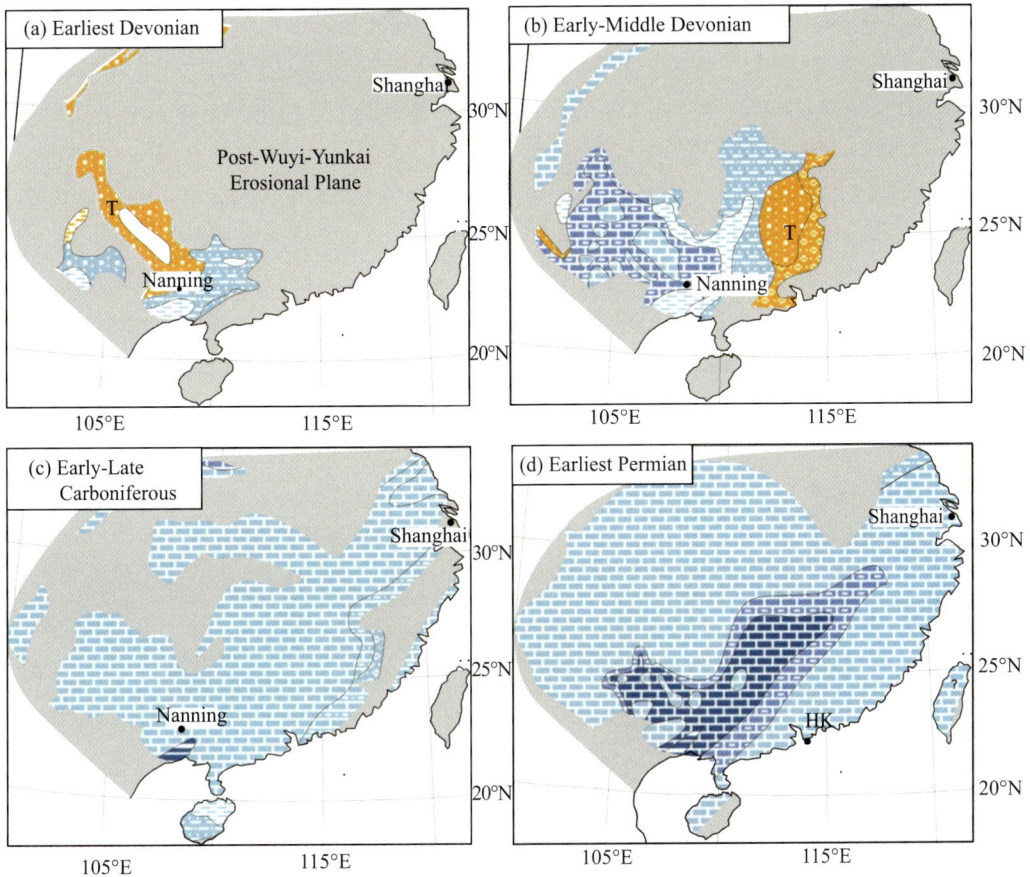

Figure 1.22   Sedimenatary facies data showing a marine transgression after the Wuyi-Yunkai Orogeny that covered the entire SCB with platform carbonates during the Carboniferous-Early Permian time (after Liu and Xu, 1994). T–terrestrial deposits

Since mid-Permian, South China became tectonically active again, having orogens developed along its NW margin (the Longmenshan Orogen; Chen and Wilson, 1996), its north margin (the Qinling-Dabie Orogen), and a ca. 1300 km-wide orogen in the southeastern > 3/5 of the continent (the South China Fold Belt or Orogen; Figure 1.24, Figure 1.25(a)). At around 260 Ma, a mantle plume (the Emeishan plume) broke out at western SCB, causing regional doming and continental flood basalts (Chung and Jahn, 1995; He et al., 2003; Xu et al., 2004; He et al., 2007). Here we will focus our discussion on the South China Orogen and the post orogenic, Jurassic–Cretaceous large magmatic province.

Most pre-2007 studies treated the orogenic event and the Jurassic–Cretaceous large magmatic province as distinctive geological events. The orogenic event, commonly referred to as the Indosinian Orogeny, is characterised by dominantly NW-directed thrusting (Hsü et al.,

1988; Wang et al., 2005; Xiao and He, 2005) and metamorphism (see review in Li and Li, 2007). The orogeny had been interpreted as either due to terrane collision (Hsü et al., 1988; Xiao and He, 2005), far-field compression (Li, 1998; Wang et al., 2007b), or Pacific subduction (Cui and Li, 1983). The terrane collision models contradict most geological observations, such as the late Paleozoic sedimentary facies data (Figure 1.22). The far-field models do not explain the geometry and kinematics of the orogen, and the simple Pacific subduction model does not explain the enormous width of the orogen. The Jurassic–Cretaceous magmatic events had commonly been regarded as a continental arc (Jahn et al., 1990), although shallow-subduction was invoked to explain the width of the "arc" magmatic belt (Zhou and Li, 2000). However, neither models account for the compositional and distributional complexities of this vast magmatic province (see discussions in Li and Li, 2007; Li et al., 2007a; Li et al., 2007b). Gilder et al. (1991) suggested that there was a Mesozoic basin-and-range province in South China, but neither the precise timing nor the cause of it was clear then.

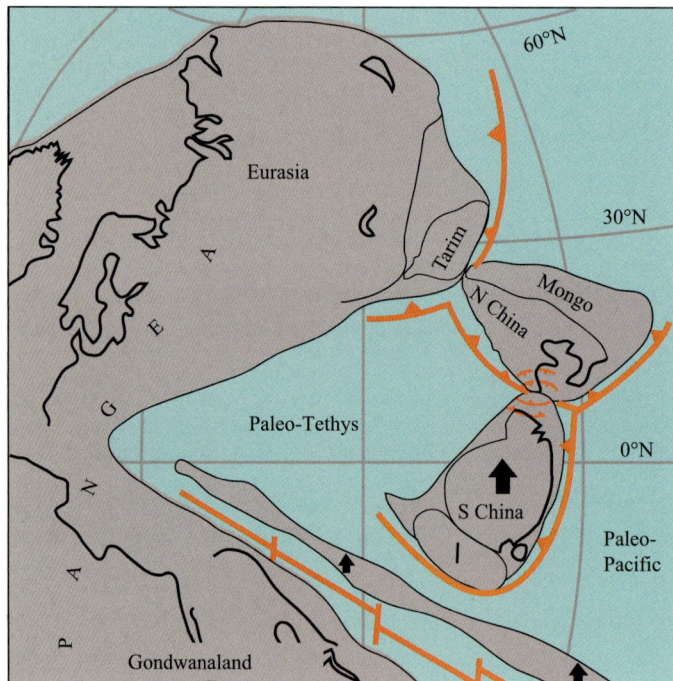

Figure 1.23 Paleogeography of mid-Permian (ca. 260 Ma) showing the collision between the North and South China blocks, and subduction along the Paleo-Pacific margin (after Li, 1998)

I—Indochina Block

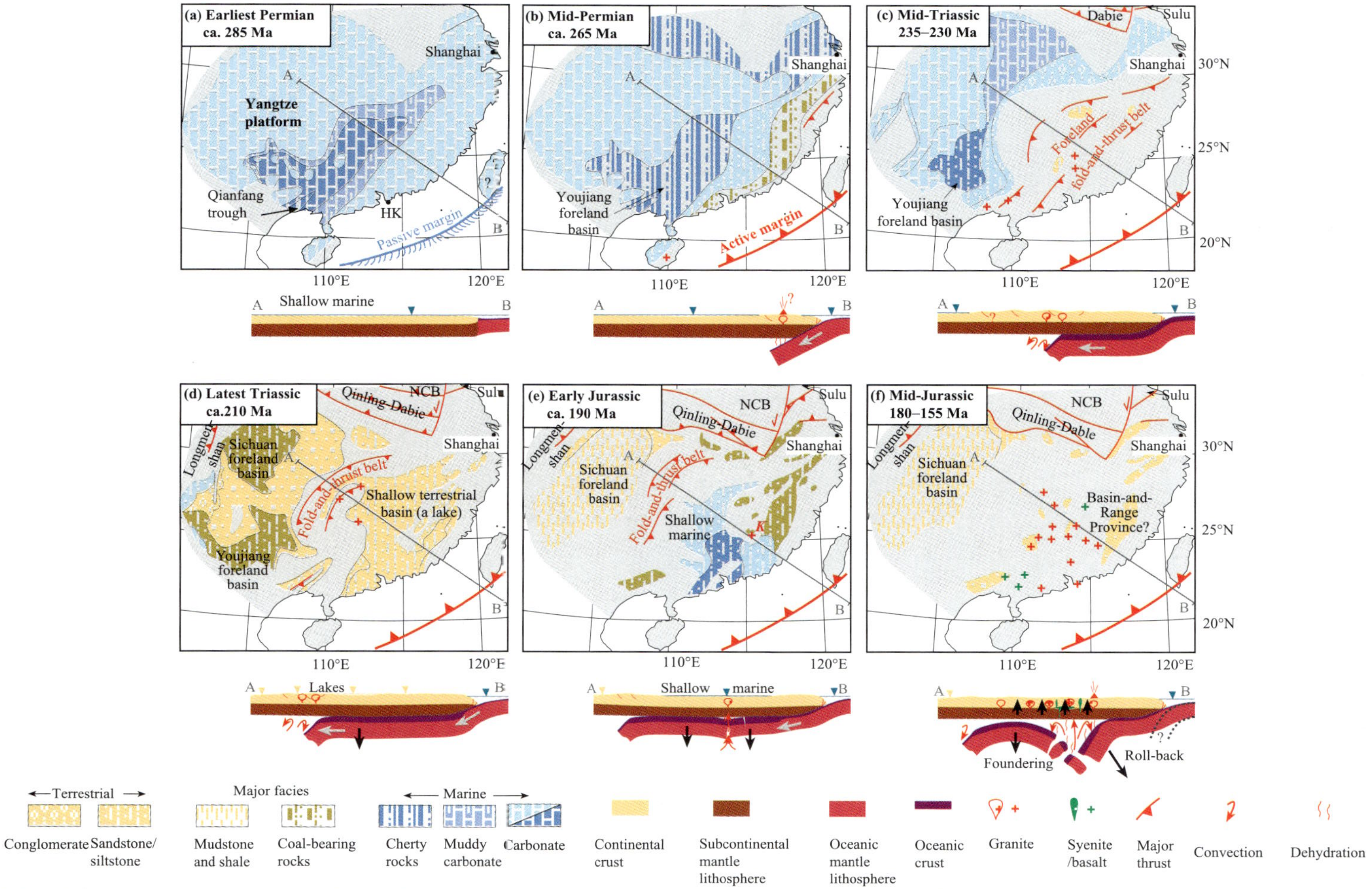

Figure 1.24 Paleogeographic evolution of the South China during the Permian–Triassic time (after Wang, 1985; Liu and Xü, 1994; Li, 1998) and the proposed flat-slab subduction model for the propagating orogen (b)–(e), the development of a sag basin at the wake of the orogen (d)–(e), and the post-orogenic magmatism due to slab foundering and a retreating arc (f). Present-day geographic grids are shown for reference. "K" marks the 189 ± 3 Ma Keshubei A-type granite (after Li and Li, 2007)

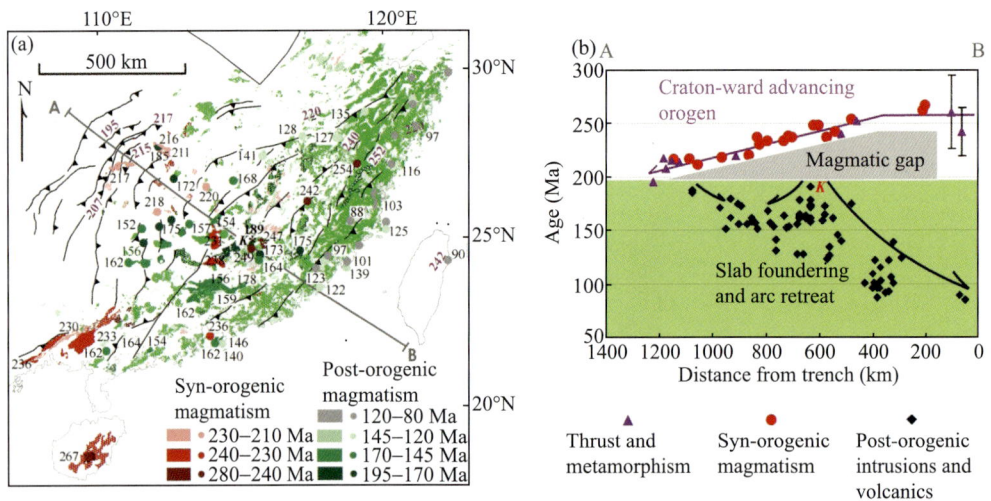

Figure 1.25   (a) Distribution of Permian–Early Cretaceous magmatism and major Mesozoic thrust faults in South China. Representative ages are shown in black for magmatic rocks and in purple for metamorphism or thrusting. Position and sense of thrusting for major thrusts were based on provincial geological memoirs, geophysical profiling (Wang, 1994), and structural mapping (e.g., Chen, 1999; Wang et al., 2005). "*K*" marks 189 ± 3 Ma Keshubei A-type granite. (b) Time-space plot of mid-Permian to mid-Cretaceous (270–80 Ma) thrusting, metamorphism, and magmatism in the South China Fold Belt as projected on cross section A–B in (a). Thrusting, metamorphic, and syn-orogenic magmatic ages were projected following structure curvatures as in (a), whereas postorogenic magmatic ages were projected directly onto the cross section

Li and Li (2007) noticed that the migration of the foreland basin, the age pattern of deformation and metamorphism, and the distribution of the limited syn-orogenic magmatism, all suggest that the Permian–Triassic orogeny propagated from the coastal region to the continental interior (Figure 1.25). This craton-ward migration pattern was also recognised by Cui and Li (1983). Li and Li (2007) also noticed that the post-orogenic magmatism exhibits more characteristics of extensional tectonics than arc-related magmatism, and that the magmatism appears to have radiated from the centre of the South China Orogen (Figure 1.25). Accompanying the orogenic and magmatic events were rapid vertical tectonic movements in the region (Li and Li, 2007) (Figure 1.24). These authors thus proposed a flat-subduction and foundering model (Figure 1.24) that includes the following major elements:

(1) Coupling between the flat-slab (a young, thick and buoyant oceanic plateau) and the overlying continental lithosphere caused the craton-ward migration of the foreland thrust belt with cessation of magmatic activity (thus a magmatic gap) above the flat-slab (Figure 1.24(b)–(d), Figure 1.25);

(2) Downward pulling of the subducted flat-slab at the rear of the migrating foreland thrust belt, due to densification by eclogite facies metamorphism led to the formation of a shal-

low-marine basin in the wake of the thrust belt (Figure 1.24(d), (e));

(3) The eventual delamination and foundering of the flat-slab from its centre led to wide-spread (radiating outward) anorogenic magmatism, lithospheric rebound and extension (Figure 1.24(f), Figure 1.25(b));

(4) Coastward retreat of the part of the delaminating slab caused the coastward migration of the Jurassic-Cretaceous magmatic belt (Figure 1.25(b)) that has mixed extensional and arc signatures.

Li Z X et al. (2012) further documented evidence for the re-initiation of normal subduction along the continental margin at around 180 Ma (dotted lines in cross-section of Figure 1.24(f)). For updated summaries of the evolution of magmatic activities in the region during the flat-slab subduction and foundering, see Meng et al. (2012), Li et al. (2013) and Zhu et al. (2013, 2014).

➢1.8   Transition from an Andean-type active margin (ca. 280 Ma to ca. 90 Ma) to the current Western Pacific-type margin soon after 90 Ma

The Permain–Cretaceous active continental margin along southeastern SCB can be regarded as a typical Andean-type convergent plate margin characterized by alternating sections of continental magmatic arc, and magmatic gaps due to flat-slab subduction, aseismic-ridge subduction or spreading ridge subduction (Ramos and Aleman, 2000) along the strike of the margin. The starting age of this continental arc system is estimated at ca. 280 Ma (Li et al., 2006a; Li X H et al., 2012). This active continental margin, producing a vast amount of magmatism in the region (Figure 1.25(a)), is in sharp contrast with the present-day Western Pacific-type plate margin where a passive continental margin is separated by young oceanic basins from oceanic arcs. Geological record shows that the transition of the Andean-type convergent plate margin to the Western Pacific-type plate margin occurred after ca. 90 Ma, roughly simultaneous along the Western Pacific margin (Li Z X et al., 2012; Figure 1.26).

Key evidences for this transition of tectonic regime include: ① the disappearance of continental arc magmatism from both East and Southeast Asia coastal regions (involving an ocean-ward jump of the subduction system; Hall, 2002), and from the eastern Australian continental margin; and ② the starting of continental rifting along the Western Pacific margin and the eventual opening of marginal seas, including the Tasman Sea, the Philippine Sea, The South China Sea, and the Sea of Japan. This dramatic transformation of tectonic regime was likely due to the roll-back of the old and heavy oceanic slabs in the Western Pacific Ocean (Schellart and Lister, 2005; Schellart et al., 2006).

Coastal magmatism

Interpreted tectonic environment

Age

0 Ma

Continental rifting leading to the opening of marginal seas,and plume magmatism around Hainan Island

South China Block

On

Western Pacific-type margin

ca. 60 Ma

Arc jumped off-shore?

A   B

Off

ca. 90 Ma

Re-initiated continental magmatic arc, +/-magmatism related to the delamination of the flat-slab

New arc

On

A   B

Andean-type margin

ca. 190 Ma

Magmatic gap due to flat-slab subduction

Off

A   B

ca. 250 Ma

Continental magmatic arc

A   B

On

ca. 280 Ma

Passive continental margin?

A Shallow marine   B

Off

Passive margin?

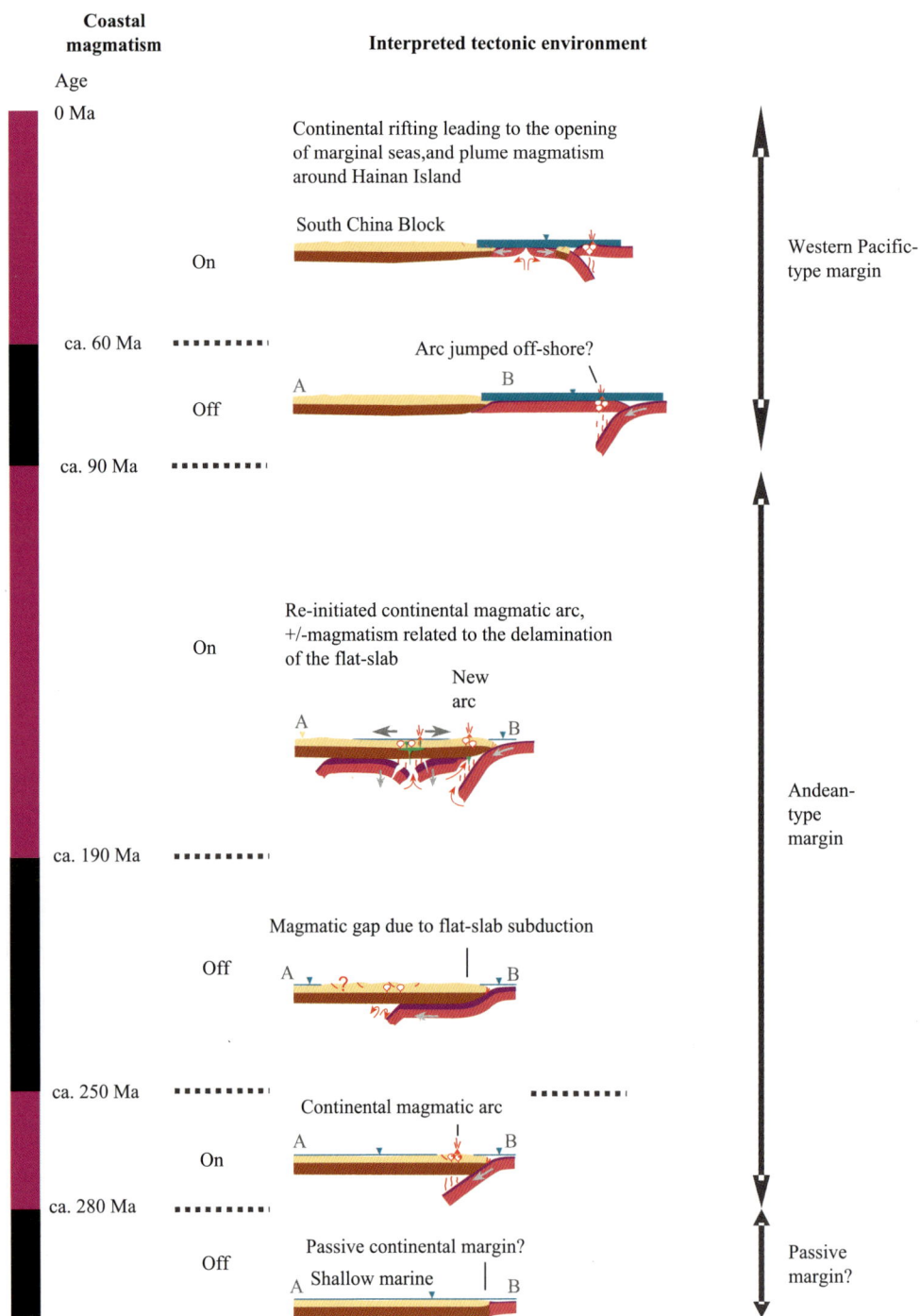

Figure 1.26   A model for the tectonic-magmatic evolution of the SCB coastal region, featuring the transition from a Permian–Cretaceous Andean-type active continental margin to a Late Cretaceous–Cenozoic Western Pacific-type plate plate margin (after Li Z X et al., 2012)

➤1.9   Effects of Cenozoic India-Eurasia collision on the western margin of the SCB

Cenozoic tectonics in the SCB was most active along its western margin, due to the continental collision between the Indo-Australian plate and Eurasia (Tapponnier et al., 1982). The dip directions of the Early Cretaceous paleomagnetic vectors suggest significant clockwise rotations occurred to the region west of the Xianshuihe-Xiaojiang Fault Zone (XXFZ in Figure 1.27), whereas anticlockwise rotations dominated within the fault zone (Figure 1.27). Some anticlockwise rotation occurred to the Hainan Island and the adjacent mainland region (Figure 1.27). However, it has been argued that the opening of the South China Sea was not a consequence of the India-Eurasia collision (Li, 2012) as proposed by some (Briais et al., 1993; Tapponnier et al., 1982); rather, it occurred as part of the back-arc extension during the ocean-ward retreat of the Western Pacific subduction system (including slab roll-back; see Section 1.7) (Schellart and Lister, 2005; Schellart et al., 2006), with contributions from the pulling force of the subducting Paleo-South China Sea slab (Figure 1.26; Holloway, 1982; Taylor and Hayes, 1983).

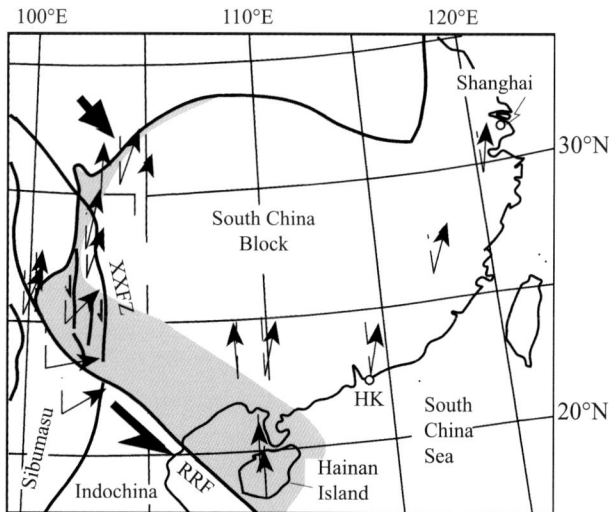

Figure 1.27   Horizontal projections of Early Cretaceous paleomagnetic directions (small arrows) in South China indicating relative block rotations along its western margin since the Cretaceous (after Li et al., 1995a).
RRF = Red River Fault; XXFZ = Xianshuihe-Xiaojiang Fault Zone

References

Anderson H E, Davis D W. 1995. U-Pb geochronology of the Moyie sills, Purcell Supergroup, southeastern British Columbia: implications for the Mesoproterozoic geological history of the Purcell (Belt) basin. Can J Earth Sci, 32: 1180–1193.

Borg S G, DePaolo D J. 1994. Laurentia, Australia, and Antarctica as a Late Proterozoic supercontinent: constraints from isotopic mapping. Geology, 22: 307–310.

Briais A, Patriat P, Tapponnier P. 1993. Updated interpretation of magnetic anomalies and seafloor spreading stages in the South China Sea: implications for the Tertiary tectonics of southeastern Asia. Journal of Geophysical Research, 98(84):6299–6328.

BOGAMR (Bureau of Geology and Mineral Resources, Jiangxi, Hunan, Fujian, Guangdong, and Guangxi). 1984. The Geological Map of Nanling and Neighbouring Area, China. Beijing: Geological Publishing House (with English explanatory notes).

Carter A, Roques D, Bristow C, et al. 2001. Understanding Mesozoic accretion in Southeast Asia: significance of Triassic thermotectonism (Indosinian Orogeny) in Vietnam. Geology, 29: 211–214.

Charvet J, Cluzel D, Faure M, et al. 1999. Some tectonic aspects of the pre-Jurassic accretionary evolution of East Asia. *In*: Metcalfe I, Ren J, Charvet J, et al. Gondwana Dispersion and Asian Accretion. Rotterdam: Balkema A A: 37–65.

Charvet J, Shu L, Faure M, et al. 2010. Structural development of the Lower Paleozoic belt of South China: genesis of an intracontinental orogen. Journal of Asian Earth Sciences, 39: 309–330.

Chen A. 1999. Mirror-image thrusting in the South China orogenic belt; tectonic evidence from western Fujian, southeastern China. Tectonophysics, 305: 497–519.

Chen C H, Lee C Y, Hsieh P S, et al. 2008. Approaching the age problem for some metamorphosed Precambrian basement rocks and Phanerozoic granitic bodies in the Wuyishan area: the application of EMP monozite age dating. Geological Journal of China Universities, 14: 1–15.

Chen J, Foland K A, Xing F, et al. 1991. Magmatism along the southeast margin of the Yangtze block: Precambrian collision of the Yangtze and Cathysia block of China. Geology, 19: 815–818.

Chen J F, Jahn B M. 1998. Crustal evolution of southeastern China: Nd and Sr isotopic evidence. Tectonophysics, 284: 101–133.

Chen S F, Wilson C J L. 1996. Emplacement of the Longmen Shan Thrust-Nappe belt along the eastern margin of the Tibetan plateau. Journal of Structural Geology, 18: 413–430.

Chen W, Harrison T M, Heizerler M T, et al. 1992. The cooling history of melange zone in north Jiangsu-south Shandong region: evidence from multiple diffusion domain ^{40}Ar-^{39}Ar thermal geochronology. Acta Petrologica Sinica, 8(1):1–17.

Chen Z, Xing G, Guo K, et al. 2009. Petrogenesis of keratophyes in the Pingshui Group, Zhejiang: Constraints from zircon U-Pb ages and Hf isotopes. Chinese Science Bulletin, 54: 1570–1578.

Chung S L, Jahn B M. 1995. Plume-lithosphere interaction in generation of the Emeishan flood basalts at the Permian-Triassic boundary. Geology, 23: 889–892.

Cui S, Li J. 1983. On the Indosinian Orogeny along the Chinese Western Pacific belt. Acta Geologica Sinica, 57(1): 51–61 (in Chinese).

Dalziel I W D. 1991. Pacific margins of Laurentia and East Antarctica-Australia as a conjugate rift pair: evidence and implications for an eocambrian supercontinent. Geology, 19: 598–601.

Davidson A. 1995. A review of the Grenville orogen in its North American type area. Journal of Australian Geology and Geophysics, 16: 3–24.

DOGNU (Department of Geology, Nanjing University). 1981. Distribution Map of Different-aged Granites in South China (1:3000000). Beijing: Science Press.

Ernst R E, Wingate M T D, Buchan K L, et al. 2008. Global record of 1600–700 Ma Large Igneous Provinces (LIPs): implications for the reconstruction of the proposed Nuna (Columbia) and Rodinia supercontinents. Precambrian Research, 160: 159–178.

Fan H P, Zhu W G, Li Z X, et al. 2013. Ca. 1.5 Ga mafic magmatism in South China during the break-up of the supercontinent Nuna/Columbia: the Zhuqing Fe-Ti-V oxide ore-bearing mafic intrusions in western Yangtze

Block. Lithos, 168–169: 85–98.

Gao S, Yang J, Zhou L, et al. 2011. Age and growth of the Archean Kongling terrain, South China, with emphasis on 3.3 Ga granitoid gneisses. American Journal of Science, 311: 153–182.

Gilder S A, Keller G R, Luo M, et al. 1991. Timing and spatial distribution of rifting in China. Tectonophysics, 197: 225–243.

Grabau A W. 1924. Stratigraphy of China. Beijing: Geological Survey of China Publication: 8–9.

Greentree M R, Li Z X. 2008. The oldest known rocks in south-western China: SHRIMP U-Pb magmatic crystallisation age and detrital provenance analysis of the Paleoproterozoic Dahongshan Group. Journal of Asian Earth Sciences, 33: 289–302.

Greentree M R, Li Z X, Li X H, et al. 2006. Late Mesoproterozoic to earliest Neoproterozoic basin record of the Sibao orogenesis in western South China and relationship to the assembly of Rodinia. Precambrian Research, 151: 79–100.

Guo L, Shi Y, Ma R, et al. 1986. The plate movement and crustal evolution of the Jiangnan Proterozoic island-arc tectonics. In: Proceedings of the International Symposium on Precambrian Crustal Evolution. Beijing: Geological Publishing House: 20–39.

Hall R. 2002. Cenozoic geological and plate tectonic evolution of SE Asia and the SW Pacific: computer-based reconstructions, model and animations. Journal of Asian Earth Sciences, 20: 353–431.

He B, Xu Y G, Chung S L, et al. 2003. Sedimentary evidence for a rapid, kilometer-scale crustal doming prior to the eruption of the Emeishan flood basalts. Earth and Planetary Science Letters, 213: 391–405.

He B, Xu Y G, Huang X L, et al. 2007. Age and duration of the Emeishan flood volcanism, SW China: Geochemistry and SHRIMP zircon U-Pb dating of silicic ignimbrites, post-volcanic Xuanwei Formation and clay tuff at the Chaotian section. Earth and Planetary Science Letters, 255: 306–323.

Herzberg C, Asimow P D, Arndt N, et al. 2007. Temperatures in ambient mantle and plumes: Constraints from basalts, picrites, and komatiites. Geochemistry, Geophysics, Geosystems, 8(2): CiteID Q02006.

Hoffman P F. 1991. Did the breakout of Laurentia turn Gondwanaland inside-out? Science, 252: 1409–1412.

Holloway N H. 1982. North Palawan block, Philippines - its relation to Asian mainland and role in evolution of South China Sea. AAPG Bulletin, 66(9):1355–1383.

Hsü K J. 1994. Tectonic facies in an archipelago model of intra-plate orogenesis. GSA Today, 4: 289–293.

Hsü K J, Sun S, Li J, et al.1988. Mesozoic overthrust tectonics in south China. Geology, 16: 418–421.

Hsü K J, Li J, Chen H, et al. 1990. Tectonics of South China: key to understanding West Pacific geology. Tectonophysics, 183: 9–39.

Hu A, Zhu B, Mao C U, et al. 1991. Geochronology of the Dahongshan Group. Chinese Journal of Geochemistry, 10: 195–203.

Huang J, Ren J, Jiang C, et al. 1980. The Geotectonic Evolution of China. Beijing:Science Press: 124.

Huang T K. 1960. The main characteristics of the geologic structures of China: preliminary conclusions. Acta Geologica Sinica, 40: 1–37 (in Chinese with extended English abstract).

Jahn B M, Zhou X H, Li J L. 1990. Formation and tectonic evolution of Southeastern China and Taiwan: Isotopic and geochemical constraints. Tectonophysics, 183: 145–160.

Jiao W F, Wu Y B, Yang S H, et al. 2009. The oldest basement rock in the Yangtze Craton revealed by zircon U-Pb age and Hf isotope composition. Science in China(Series D), 52: 1393–1399.

Kim S W, Oh C W, Williams I S, et al. 2006. Phanerozoic high-pressure eclogite and intermediate-pressure granulite facies metamorphism in the Gyeonggi Massif, South Korea: implications for the eastward extension of the Dabie-Sulu continental collision zone. Lithos, 92: 357–377.

Li J. 1992. Tectonic framework and evolution of southeastern China. J SE Asian Earth Sci, 8: 219–223.

Li W X, Li X H. 2003. Adakitic granites within the NE Jiangxi ophiolites, South China: geochemical and Nd isotopic evidence. Precambrian Research, 122: 29–44.

Li W X, Li X H, Li Z X. 2005. Neoproterozoic bimodal magmatism in the Cathaysia Block of South China and its tectonic significance. Precambrian Research, 136: 51–66.

Li W X, Li X H, Li Z X, et al. 2008a. Obduction-type granites within the NE Jiangxi Ophiolite: implications for the final amalgamation between the Yangtze and Cathaysia Blocks. Gondwana Research, 13: 288–301.

Li W X, Li X H, Li Z X. 2008b. Middle Neoproterozoic syn-rifting volcanic rocks in Guangfeng, South China: petrogenesis and tectonic significance. Geological Magzine, 145: 475–489.

Li W X, Li X H, Li Z X. 2010. Ca. 850 Ma bimodal volcanic rocks in northeastern Jiangxi Province, South China: initial extension during the breakup of Rodinia? American Journal of Science, 310: 951–980.

Li X H. 1991. Geochronology of the Wanyangshan-Zhuguangshan granitoid batholith: Implication for the crust development. Science in China (Series B), 34: 620–629.

Li X H. 1997. Timing of the Cathaysia Block formation: constraints from SHRIMP U-Pb zircon geochronology. Episodes, 20: 188–192.

Li X H. 1999. U-Pb zircon ages of granites from the southern margin of the Yangtze Block: timing of the Neoproterozoic Jinning Orogeny in SE China and implications for Rodinia assembly. Precambrian Research, 97: 43–57.

Li X H, Gui X T. 1991. Source rock of the Caledonian-age granitoid rocks in Wanyangshan, southeast China, I. Science in China (Series B), 5: 533–540.

Li X H, Tatsumoto M, Premo W R, et al. 1989. Age and origin of the Tanghu granite, southeast China − results from U-Pb single zircon and Nd isotopes. Geology, 17: 395–399.

Li X H, Zhou G, Zhao J, et al. 1994. SHRIMP ion microprobe zircon U-Pb age of the NE Jiangxi ophiolite and its tectonic implications. Geochimica, 23: 125–131.

Li X H, Li Z X, Zhou H, et al. 2002. U-Pb zircon geochronology, geochemistry and Nd isotopic study of Neoproterozoic bimodal volcanic rocks in the Kangdian Rift of South China: implications for the initial rifting of Rodinia. Precambrian Research, 113: 135–154.

Li X H, Li Z X, Ge W C, et al. 2003a. Neoproterozoic granitoids in South China: crustal melting above a mantle plume at ca. 825 Ma? Precambrian Research, 122: 45–83.

Li X H, Li Z X, Zhou H, et al. 2003b. SHRIMP U-Pb zircon age, geochemistry and Nd isotope of the Guandaoshan pluton in SW Sichuan: petrogenesis and tectonic significance. Science in China(Series D), 46 (Supp): 73–83.

Li X H, Li Z X, Li W X, et al. 2006a. Initiation of the Indosinian Orogeny in South China: evidence for a Permian magmatic arc on Hainan Island. Journal of Geology, 114: 341–353.

Li X H, Li Z X, Sinclair J A, et al. 2006b. Revisiting the "Yanbian Terrane": implications for Neoproterozoic tectonic evolution of the western Yangtze Block, South China. Precambrian Research, 151: 14–30.

Li X H, Li W X, Li Z X. 2007a. On the genetic classification and tectonic implications of the Early Yanshanian granitoids in the Nanling Range, South China. Chinese Science Bulletin, 52:1873–1885.

Li X H, Li Z X, Li W X, et al. 2007b. U-Pb zircon, geochemical and Sr-Nd-Hf isotopic constraints on age and origin of Jurassic I- and A-type granites from central Guangdong, SE China: a major igneous event in response to foundering of a subducted flat-slab? Lithos, 96: 186–204.

Li X H, Li Z X, Sinclair J A, et al. 2007c. Reply to the comment by Zhou et al. on: "Revisiting the 'Yanbian Terrane': implications for Neoproterozoic tectonic evolution of the western Yangtze Block, South China" - [Precambrian Research 151 (2006) 14–30]-[Precambrian Research 154 (2007) 153–157]. Precambrian Research, 155: 318–323.

Li X H, Li W X, Li Z X, et al. 2008. 850–790 Ma bimodal volcanic and intrusive rocks in northern Zhejiang, South

China: a major episode of continental rift magmatism during the breakup of Rodinia. Lithos, 102: 341–357.

Li X H, Li W X, Li Z X, et al. 2009. Amalgamation between the Yangtze and Cathaysia Blocks in South China: constraints from SHRIMP U-Pb zircon ages, geochemistry and Nd-Hf isotopes of the Shuangxiwu volcanic rocks. Precambrian Research, 174: 117–128.

Li X H, Li W X, Li Q L, et al. 2010. Petrogenesis and tectonic significance of the ~850 Ma Gangbian alkaline complex in South China: evidence from in situ zircon U-Pb dating, Hf-O isotopes and whole-rock geochemistry. Lithos, 114: 1–15.

Li X H, Li Z X, He B, et al. 2012. The Early Permian active continental margin and crustal growth of the Cathaysia Block: in situ U–Pb, Lu–Hf and O isotope analyses of detrital zircons. Chemical Geology, 328: 195–207.

Li X H, Li Z X, Li W X, et al. 2013. Revisiting the "C-type adakites" of the Lower Yangtze River Belt, central eastern China: In-situ zircon Hf–O isotope and geochemical constraints. Chemical Geology, 345: 1–15.

Li Z X. 1998. Tectonic history of the major East Asian lithospheric blocks since the Mid-Proterozoic—a synthesis. *In*: Flower M J, Chung S L, Lo C H, et al. Mantle Dynamics and Plate Interactions in East Asia. Washington D C: American Geophysical Union: 221–243.

Li Z X. 2012. The opening of the South China Sea: driven by Pacific subduction, or by India-Eurasia collision? AGU Fall Meeting. American Geophysical Union, San Francisco, abstract, T42A–41.

Li Z X, Li X H. 2007. Formation of the 1300-km-wide intracontinental orogen and postorogenic magmatic province in Mesozoic South China: a flat-slab subduction model. Geology, 35 179–182.

Li Z X, Metcalfe I, Wang X. 1995a. Vertical-axis block rotations in southwestern China since the Cretaceous: new paleomagnetic results from Hainan Island. Geophysics Research Letters, 22: 3071–3074.

Li Z X, Zhang L H, Powell C M. 1995b. South China in Rodinia: part of the missing link between Australia-East Antarctica and Laurentia. Geology, 23: 407–410.

Li Z X, Li X H, Kinny P D, et al. 1999. The breakup of Rodinia: did it start with a mantle plume beneath South China? Earth and Planetary Science Letters, 173: 171–181.

Li Z X, Li X H, Zhou H W, et al. 2002. Grenvillian continental collision in South China: New SHRIMP U-Pb zircon results and implications for the configuration of Rodinia. Geology, 30: 163–166.

Li Z X, Li X H, Kinny P D, et al. 2003a. Geochronology of Neoproterozoic syn-rift magmatism in the Yangtze Craton, South China and correlations with other continents: evidence for a mantle superplume that broke up Rodinia. Precambrian Research, 122: 85–109.

Li Z X, Li X H, Wang J, et al. 2003b. A tectonic overview of the South China Block. *In*: Li Z X, Wang J, Li X H, et al. From Sibao Orogenesis to Nanhua Rifting: Late Precambrian Tectonic History of Eastern South China. Beijing: Geological Publishing House: 1–13.

Li Z X, Evans D A D, Zhang S. 2004. A 90° spin on Rodinia: Possible causal links between the Neoproterozoic supercontinent, superplume, true polar wander and low-latitude glaciation. Earth and Planetary Science Letters, 220: 409–421.

Li Z X, Wartho J A, Occhipinti S, et al. 2007. Early history of the eastern Sibao Orogen (South China) during the assembly of Rodinia: New mica [40]Ar/[39]Ar dating and SHRIMP U-Pb detrital zircon provenance constraints. Precambrian Research, 159: 79–94.

Li Z X, Li X H, Li W, et al. 2008a. Was Cathaysia part of Proterozoic Laurentia? new data from Hainan Island, south China. Terra Nova, 20: 154–164.

Li Z X, Bogdanova S V, Collins A S, et al. 2008b. Assembly, configuration, and break-up history of Rodinia: a synthesis. Precambrian Research, 160: 179–210.

Li Z X, Li X H, Wartho J A, et al. 2010. Magmatic and metamorphic events during the Early Paleozoic Wuyi-Yunkai Orogeny, southeastern South China: new age constraints and *P-T* conditions. Geological Society of

America Bulletin, 122: 772–793.

Li Z X, Li X H, Chung S L, et al. 2012. Magmatic switch-on and switch-off along the South China continental margin since the Permian: transition from an Andean-type to a Western Pacific-type plate boundary. Tectonophysics, 532-535: 271–290.

Lin G C, Li X H, Li W X. 2007. SHRIMP U-Pb zircon age, geochemistry and Nd-Hf isotope of Neoproterozoic mafic dyke swarms in western Sichuan: petrogenesis and tectonic significance. Science in China(Series D), 50: 1–16.

Ling W L, Gao S, Zhang B R, et al. 2003. Neoproterozoic tectonic evolution of the northwestern Yangtze craton, South China: implications for amalgamation and break-up of the Rodinia Supercontinent. Precambrian Research, 122: 111–140.

Liu B, Xü X. 1994. Atlas of Lithofacies and Paleogeography of South China. Beijing: Science Press: 188.

Liu H. 1991. Correlation of Sinian system. Beijing:Science Press:126–170.

Ma D, Huang X, Xiao Z, et al. 1998. The crystalline basement of the Hainan Island—stratigraphy and geochronology of the Baoban Group. Wuhan: China University of Geosciences Press: 60(in Chinese).

Meng L F, Li Z X, Chen H L, et al. 2012. Geochronological and geochemical results from Mesozoic basalts in southern South China Block support the flat-slab subduction model. Lithos, 132–133: 127–140.

Moores E M. 1991. Southwest U S-East Antarctic (SWEAT) connection: a hypothesis. Geology, 19: 425–428.

Nagy E A, Maluski H, Lepvrier C, et al. 2001. Geodynamic significance of the Kontum massif in central Vietnam: composite $^{40}Ar/^{39}Ar$ and U-Pb ages from Paleozoic to Triassic. Journal of Geology, 109: 755–770.

Qiao X, Geng S. 1981. On late Precambrian plate tectonics of South China. In: Huang J, Li C. Contributions to the Tectonics of China and Adjacent Regions. Beijing:Geological Publishing House: 77–91.

Qiu Y M, Gao S, McNaughton N J, et al. 2000. First evidence of >3.2 Ga continental crust in the Yangtze craton of South China and its implications for Archean crustal evolution and Phanerozoic tectonics. Geology, 28: 11–14.

Ramos V A, Aleman A. 2000. Tectonic evolution of the Andes. In: Cordani U G, Thomaz F A, Campos D A. Tectonic Evolution of South America. 31st International Geological Congress, Rio de Janeiro, Brazil: 635–688.

Ren J S. 1964. A preliminary study on pre-Devonian geotectonic problems of southeastern China. Acta Geol Sinica, 44: 418–431.

Ren J S. 1990. On the geotectonics of southeastern China. Acta Geologica Sinica, 64(2): 275–288.

Ren J S. 1991. On the geotectonics of southern China. Acta Geologica Sinica,4: 111–130.

Ren J S, Wang Z, Chen B, et al. 1997. Tectonic Map of China and Adjacent Regions. Beijing:Geological Publishing House.

Rivers T. 1997. Lithotectonic elements of the Grenville Province: review and tectonic implications. Precambrian Research, 86:117–154.

Roger F, Maluski H, Leyreloup A, et al. 2007. U-Pb dating of high temperature metamorphic episodes in the Kon Tum Massif (Vietnam). Journal of Asian Earth Sciences, 30: 565–572.

Ross G M, Parrish R R, Winston D. 1992. Provenance and U-Pb geochronology of the Mesoproterozoic Belt Supergroup (northwestern United States): implications for age of deposition and pre-Panthalassa plate reconstructions. Earth and Planetary Science Letters, 113: 57–76.

Schellart W P, Lister G S. 2005. The role of the East Asian active margin in widespread extensional and strike-slip deformation in East Asia. Journal of Geology Society of London, 162: 959–972.

Schellart W P, Lister G S, Toy V G. 2006. A Late Cretaceous and Cenozoic reconstruction of the Southwest Pacific region: tectonics controlled by subduction and slab rollback processes. Earth-Science Reviews, 76: 191–233.

Shu L S, Faure M, Yu J H, et al. 2011. Geochronological and geochemical features of the Cathaysia block (South China): new evidence for the Neoproterozoic breakup of Rodinia. Precambrian Research, 187: 263–276.

Shui T. 1987. Tectonic framework of the southeastern China continental basement. Scientia Sinica, B30: 414–422.

Shui T, Xu B, Liang R, et al. 1988. Metamorphic Basement Geology of the Zhe-Min Region, China. Beijing:Science Press: 85(plus 30 plates; in Chinese).

Sinclair J A. 2001. Petrology, geochemistry, and geochronology of the "Yanbian ophiolite suite", South China: implications for the western extension of the Sibao Orogen. Honours thesis, Department of Geology and Geophysics, The University of Western Australia, Perth(plus appendixes).

Sun Z M, Yin F G, Guan J L, et al. 2009. SHRIMP U-Pb dating and its stratigraphic significance of tuff zircons from Heishan formation of Kunyang Group, Dongchuan area, Yunnan Province, China. Geological Bulletin of China, 28: 896–900 (in Chinese with English abstract).

Sun W H, Zhou M F, Gao J F, et al. 2009. Detrital zircon U-Pb geochronological and Lu-Hf isotopic constraints on the Precambrian magmatic and crustal evolution of the western Yangtze Block, SW China. Precambrian Research, 172: 99–126.

Tapponnier P, Peltzer G, Dain A Y L,et al. 1982. Propagation extrusion tectonics in Asia: new insights from simple experiments with plasticine. Geology, 611–616.

Taylor B, Hayes D E. 1983. Origin and history of the South China Sea. *In*: Hayes D E. The Tectonic and Geologic Evolution of Southeast Asian Seas and Islands, Part 2. American Geophysical Union, Washington D C:23–56.

Wan Y, Liu D, Xu M, et al. 2007. SHRIMP U-Pb zircon geochronology and geochemistry of metavolcanic and metasedimentary rocks in Northwestern Fujian, Cathaysia block, China: tectonic implications and the need to redefine lithostratigraphic units. Gondwana Research, 12: 166–183.

Wang H. 1985. Atlas of the Palaeogeography of China. Beijing:Cartographic Publishing House: 281.

Wang H, Mo X. 1995. An outline of the tectonic evolution of China. Episodes, 18: 6–16.

Wang J, Li Z X. 2003. History of Neoproterozoic rift basins in South China: implications for Rodinia break-up. Precambrian Research, 122: 141–158.

Wang J, Sun D, Chang X, et al. 1998. U-Pb dating of the Napeng granite at the NW margin of the Yunkai Block, Guangdong, South China. Acta Mineralogica Sinica, 18: 130–133.

Wang L G, Luo Z K, McNaughton N J, et al. 1996. SHRIMP U-Pd in zircon studies of plutonic rocks from the Jiaodong Peninsula, Shandong Province, China: constraints on crustal and tectonic evolution and genesis of gold mineralization. 30th International Geological Convention Abstract. 2/3: 620.

Wang M J. 1994. Gravity and magnetic interpretation of Heishui-Quanzhou geoscience transect. Acta Geologica Sinica, 37: 321–329 (in Chinese with English abstract)..

Wang Q, Wyman D A, Li Z X, et al. 2010. Petrology, geochronology and geochemistry of ca. 780 Ma A-type granites in South China: petrogenesis and implications for crustal growth during the breakup of the supercontinent Rodinia. precambrian Research, 178: 185–208.

Wang X C, Li X H, Li W X, et al. 2009. Variable involvements of mantle plumes in the genesis of mid-Neoproterozoic basaltic rocks in South China: a review. Gondwana Research, 15: 381–395.

Wang X C, Li X H, Li W X, et al. 2008. The Bikou basalts in the northwestern Yangtze block, South China: remnants of 820–810 Ma continental flood basalts. Geological Society of America Bulletin, 120: 1478–1492.

Wang X C, Li X H, Li W X, et al. 2007. Ca. 825 Ma komatiitic basalts in South China: first evidence for > 1500 degrees C mantle melts by a Rodinian mantle plume. Geology, 35: 1103–1106.

Wang X L, Zhou J C, Qiu J S, et al. 2006. LA-ICP-MS U-Pb zircon geochronology of the Neoproterozoic igneous rocks from Northern Guangxi, South China: implications for tectonic evolution. Precambrian Research, 145: 111–130.

Wang Y J, Fan W, Cawood P A, et al. 2007b. Indosinian high-strain deformation for the Yunkaidashan tectonic belt, south China: Kinematics and Ar^{40}/Ar^{39} geochronological constraints. Tectonics, 26 doi: 1029/2007 TC002099.

Wang Y J, Fan W M, Zhao G C, et al. 2007a. Zircon U-Pb geochronology of gneissic rocks in the Yunkai massif and its implications on the Caledonian event in the South China Block. Gondwana Research, 12: 404–416.

Wang Y J, Zhang Y H, Fan W M, et al. 2005. Structural signatures and $^{40}Ar/^{39}Ar$ geochronology of the Indosinian Xuefengshan tectonic belt, South China Block. Journal of Structural Geology, 27: 985–998.

Wu R X, Zheng Y F, Wu Y B, et al. 2006. Reworking of juvenile crust: element and isotope evidence from Neoproterozoic granodiorite in South China. Precambrian Research, 146: 179–212.

Xiang H, Zhang L, Zhou H W, et al. 2008. U-Pb zircon geochronology and Hf isotope study of metamorphosed basic-ultrabasic rocks from metamorphic basement in southwestern Zhejiang: the response of the Cathaysia Block to Indosinian orogenic event. Science in China(Series D), 51: 788–800.

Xiao W, He H. 2005. Early Mesozoic thrust tectonics of the northwest Zhejiang region (Southeast China). Geological Society of America Bulletin, 117: 945–961.

Xu X, O'Reilly S Y, Griffin W L, et al. 2005. Relict Proterozoic basement in the Nanling Mountains (SE China) and its tectonothermal overprinting. Tectonics, 24: doi:10.1029/2004TC001652.

Xu X S, O'Reilly S Y, Griffin W L, et al. 2007. The crust of Cathaysia: age, assembly and reworking of two terranes. Precambrian Research, 158: 51–78.

Xu Y G, He B, Chung S L, et al. 2004. Geologic, geochemical, and geophysical consequences of plume involvement in the Emeishan flood-basalt province. Geology, 32: 917–920.

Yao W H, Li Z X, Li W X, et al. 2012. Post-kinematic lithospheric delamination of the Wuyi–Yunkai orogen in South China: evidence from ca. 435 Ma high-Mg basalts. Lithos, 154: 115–129.

Yao W H, Li Z X, Li W X, et al. 2014. From Rodinia to Gondwanaland: A tale of detrital zircon provenance analyses from the southern Nanhua Basin, South China. American Journal of Science, 314: 278–313.

Ye M F, Li X H, Li W X, et al. 2007. SHRIMP zircon U-Pb geochronological and whole-rock geochemical evidence for an early Neoproterozoic Sibaoan magmatic arc along the southeastern margin of the Yangtze Block. Gondwana Research, 12: 144–156.

Yu J H, Wang L, O'Reilly S Y, et al. 2009. A Paleoproterozoic orogeny recorded in a long-lived cratonic remnant (Wuyishan terrane), eastern Cathaysia Block, China. Precambrian Research, 174: 347–363.

Zeng W, Zhang L, Zhou H W, et al. 2008. Caledonian reworking of Paleoproterozoic basement in the Cathaysia Block: Constraints from zircon U-Pb dating, Hf isotopes and trace elements. Chinese Science Bulletin, 53: 895–904.

Zhang S B, Zheng Y F, Wu Y B, et al. 2006. Zircon isotope evidence for ≥3.5 Ga continental crust in the Yangtze craton of China. Precambrian Research, 146: 16–34.

Zhao G C, Cawood P A. 1999. Tectonothermal evolution of the Mayuan assemblage in the Cathaysia Block: implications for neoproterozoic collision- related assembly of the South China craton. American Journal of Science, 299: 309–339.

Zhao X, Coe R S. 1987. Palaeomagnetic constraints on the collision and rotation of North and South China. Nature, 327: 141–144.

Zhao X F, Zhou M F, Li J W, et al. 2010. Late Paleoproterozoic to Early Mesoproterozoic Dongchuan Group in Yunnan, SW China: implications for tectonic evolution of the Yangtze Block. Precambrian Research, 182: 57–69.

Zhou G. 1989. The discovery and significance of the northeastern Jiangxi Province ophiolite (NEJXO), its metamorphic peridotite and associated high temperature-high pressure metamorphic rocks. Journal of Southeast Asian Earth Sciences, 3: 237–247.

Zhou J C, Wang X L, Qiu J S. 2009. Geochronology of Neoproterozoic mafic rocks and sandstones from northeastern Guizhou, South China: coeval arc magmatism and sedimentation. Precambrian Research, 170:

27–42.

Zhou J, Li X H, Ge W, et al. 2007. Age and origin of Middle Neoproterozoic mafic magmatism in southern Yangtze Block and relevance to the break-up of Rodinia. Gondwana Research, 12: 184–197.

Zhou M F, Kennedy A K, Sun M, et al. 2002a. Neoproterozoic arc-related mafic intrusions along the northern margin of South China: implications for the accretion of Rodinia. Journal of Geology, 110: 611–618.

Zhou M F, Yan D P, Kennedy A K, et al. 2002b. SHRIMP U-Pb zircon geochronological and geochemical evidence for Neoproterozoic arc-magmatism along the western margin of the Yangtze Block, South China. Earth and Planetary Science Letters, 196: 51–67.

Zhou M F, Kennedy A K, Sun M, et al. 2002c. Neoproterozoic arc-related mafic intrusions along the northern margin of South China: implications for the accretion of Rodinia. Journal of Geology, 110: 611–618.

Zhou M F, Ma Y X, Yan D P, et al. 2006. The Yanbian terrane (Southern Sichuan Province, SW China): a neoproterozoic arc assemblage in the western margin of the Yangtze block. Precambrian Research, 144: 19–38.

Zhou X M, Li W X. 2000. Origin of Late Mesozoic igneous rocks in Southeastern China: implications for lithosphere subduction and underplating of mafic magmas. Tectonophysics, 326: 269–287.

Zhu G, Evans J A, Fitches W R, et al.1994. Isotopic constraints on the Palaeozoic evolution of the Shandong Peninsula. Journal of Southeast Asian Earth Sciences, 9: 241–248.

Zhu K Y, Li Z X, Xu X S, et al. 2013. Late Triassic melting of a thickened crust in southeastern China: evidence for flat-slab subduction of the Paleo-Pacific plate. Journal of Asian Earth Sciences, 74: 265–279.

Zhu K Y, Li Z X, Xu X S, et al. 2014. A Mesozoic Andean-type orogenic cycle in southeastern China as recorded by granitoid evolution. American Journal of Science, 314: 187–234.

Zhu W G, Zhong H, Deng H L, et al. 2006. SHRIMP zircon U-Pb age, geochemistry, and Nd-Sr isotopes of the Gaojiacun mafic-ultramafic intrusive complex, Southwest China. International Geological Review, 48: 650–668.

PART **2**

A TOUR OF THE GEOLOGICAL HISTORY OF EASTERN SOUTH CHINA IN EIGHT DAYS[*]

Aim of the trip

To provide participants first-hand knowledge of rock outcrops that recorded major geological events that shaped the present composition and configuration of eastern South China, and sharpen their skills in interpreting geological processes based on field and analytical information

Audience

Anyone who has a geoscience degree, and has interests in reading the rock record and understanding the geological history of south China

Caution

People say that we only see what we know or what we think we know, but they forgot that rocks can also teach us what we did not know

What we will see will by no means be complete, and the interpretations may not all be correct either — one is expected to see something new every time he/she revisits the same outcrop

A challenge to all

See something new, come up with more robust interpretations, develop new hypotheses and design ways to test them

Magnetic compass correction

Magnetic declination of the region is between 4.2° and 5.1° west of true north in July 2013, changing by 0.05° west each year

(for update see http://www.ngdc.noaa.gov/geomag-web/#declination)

---

[*] Descriptions of some field stops were modified from that of Li et al. (2003b).

Unless otherwise acknowledged, all photos used in the figures were taken by Zheng-Xiang Li.

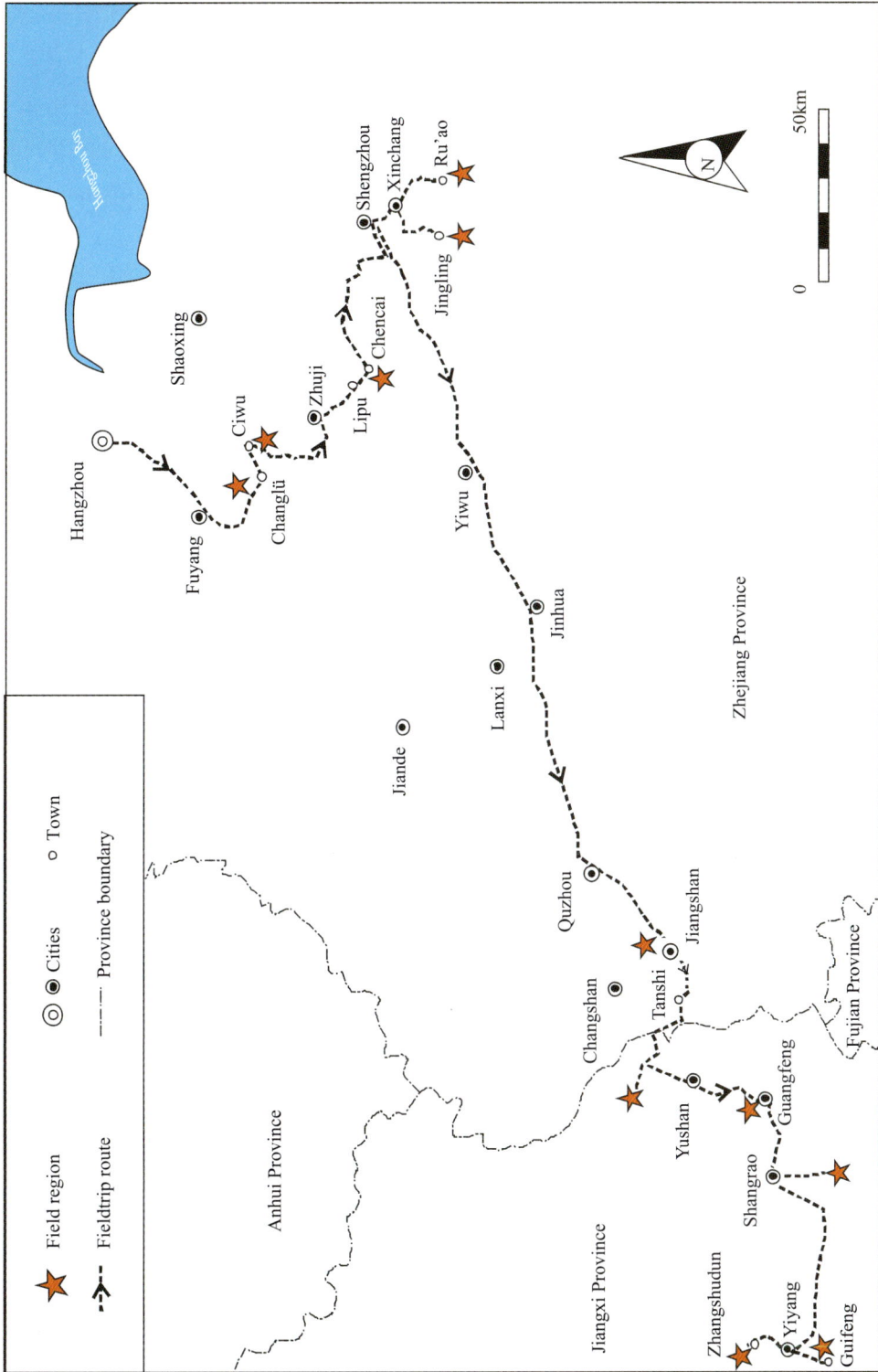

➢Day 1. Sibao orogenesis to Neoproterozoic rift magmatism

Aim: We will examine the change of tectonic regime in the South China Block from late Mesoproterozoic—earliest Neoproterozoic convergence (traditionally called the Shengong Orogeny here, regionally known as the Sibao or Jiangnan Orogeny) to mid-Neoproterozoic extension (Nanhua rifting).

Contents: The Shuangxiwu arc volcaniclastic succession, angular unconformity above it that reflects the Shengong (Sibao) Orogeny, basal rift succession, and 850–780 Ma bimodal intrusions.

Early morning departure from Zhejiang University to the Shuangxiwu region, Fuyang City.

Stop 1.1 Shengong unconformity at Shengong (now Qingong) village (29°53.302′N, 120°02.316′E)

At this stop, we will examine the contact relationship between the 970–890 Ma Shuangxiwu arc (Li et al., 2009) and the Neoproterozoic Luojiamen Formaion (Figure 2.2). The basal conglomerate of the Luojiamen Formation was deposited unconformably over an uneven paleo-erosion surface with variable compositions. This unconformity is a key evidence for the Sibao Orogeny (locally called the Shengong Orogeny, named after the Shengong village) in the region, interpreted as reflecting the transition from tectonic convergence to continental rifting during the early Neoproterozoic (Figure 1.3).

Gravel to pebble clasts in the basal conglomerate are mostly 2 to 10 cm in dimensions with a maximum size of ca. 50 cm, commonly poorly sorted and subangular-subrounded. They include granite, basement metamorphic rocks and volcanic clasts (Figure 2.3). Beneath the unconformity is the strongly cleaved Zhangcun Formation which is a tuff. Both the bedding of the Zhangcun Formation, as shown by the compositional bands, and the strong cleavage, are sub-vertical. Note that this NE-striking cleavage does not affect the overlying Neoproterozoic Luojiamen Formation, which has a much shallower bedding dip (ca. 20° to NW in the village).

The Luojiamen Formation is the basal unit of the Neoproterozoic Heshangzhen Group, interpreted as a continental rift succession (see reviews by Wang and Li, 2003). It displays an overall fining-upward trend. Across the Shengong village, we could see outcrops of a ca. 30 m thick tuffaceous breccia in lower Luojiamen Formation. Further up the succession, the rocks become dominated by tuffaceous sandstones. An unpublished SHRIMP analysis on the tuffaceous breccia revealed zircon ages of between 806 Ma and 922 Ma, reflecting a likely ca. 800 Ma deposition age and strong inheritance of country rocks (as quoted in Li et al., 2003b). SHRIMP dating of bimodal volcanic rocks in the overlying Hongchicun and Shangshu forma-

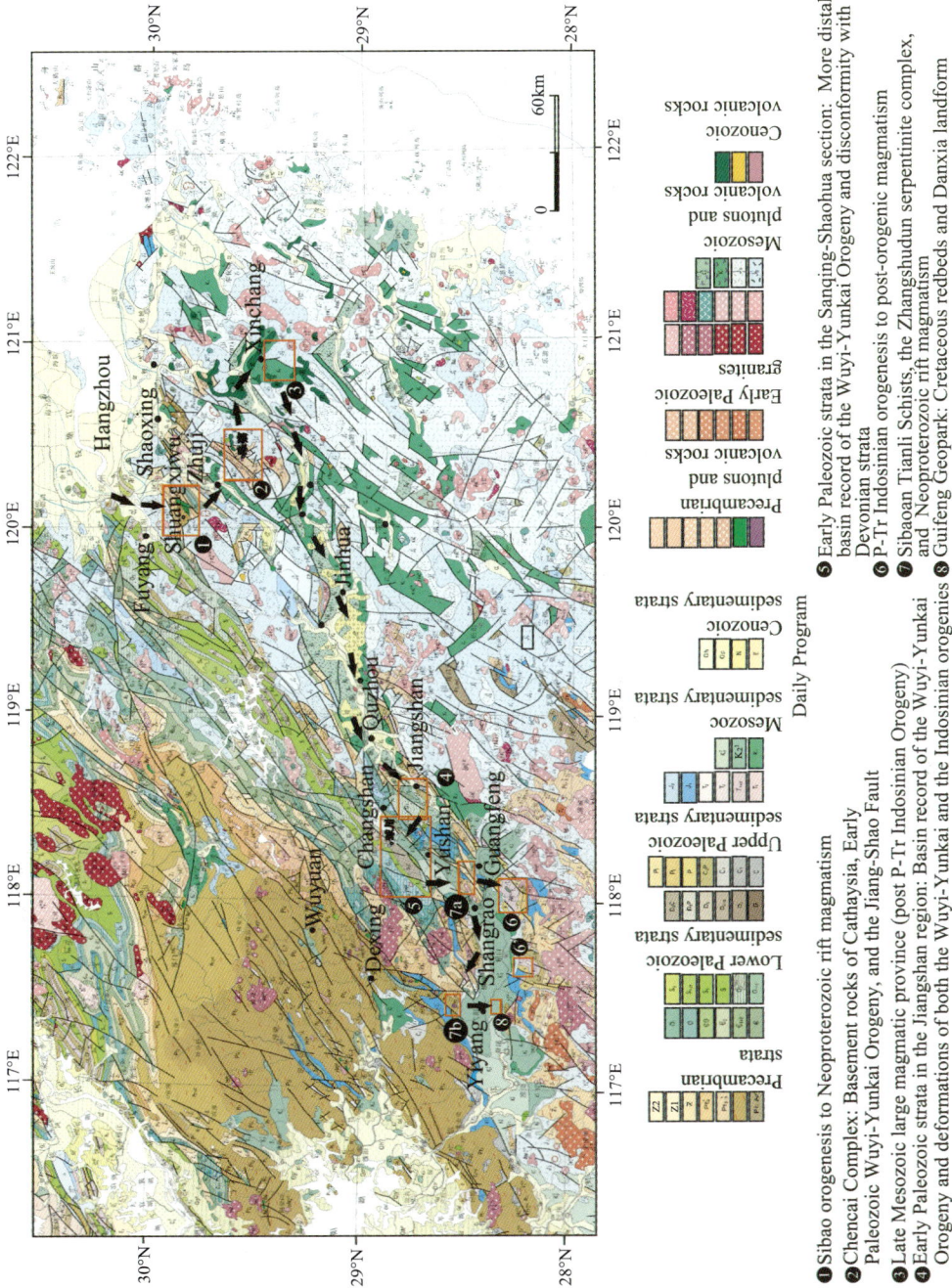

Figure 2.1 Locations of major field stops and the geological contents. The background geological map is from the 1:1,500,000 geological map of Zhejiang, Jiangxi and Fujian provinces (Ma et al., 2002)

❶ Sibao orogenesis to Neoproterozoic rift magmatism
❷ Chencai Complex: Basement rocks of Cathaysia, Early Paleozoic Wuyi-Yunkai Orogeny, and the Jiang-Shao Fault
❸ Late Mesozoic large magmatic province (post P-Tr Indosinian Orogeny)
❹ Early Paleozoic strata in the Jiangshan region: Basin record of the Wuyi-Yunkai Orogeny and deformations of both the Wuyi-Yunkai and the Indosinian orogenies
❺ Early Paleozoic strata in the Sanqing-Shaohua section: More distal basin record of the Wuyi-Yunkai Orogeny and disconformity with Devonian strata
❻ P-Tr Indosinian orogenesis to post-orogenic magmatism
❼ Sibaoan Tianli Schists, the Zhangshudun serpentinite complex, and Neoproterozoic rift magmatism
❽ Guifeng Geopark: Cretaceous redbeds and Danxia landform

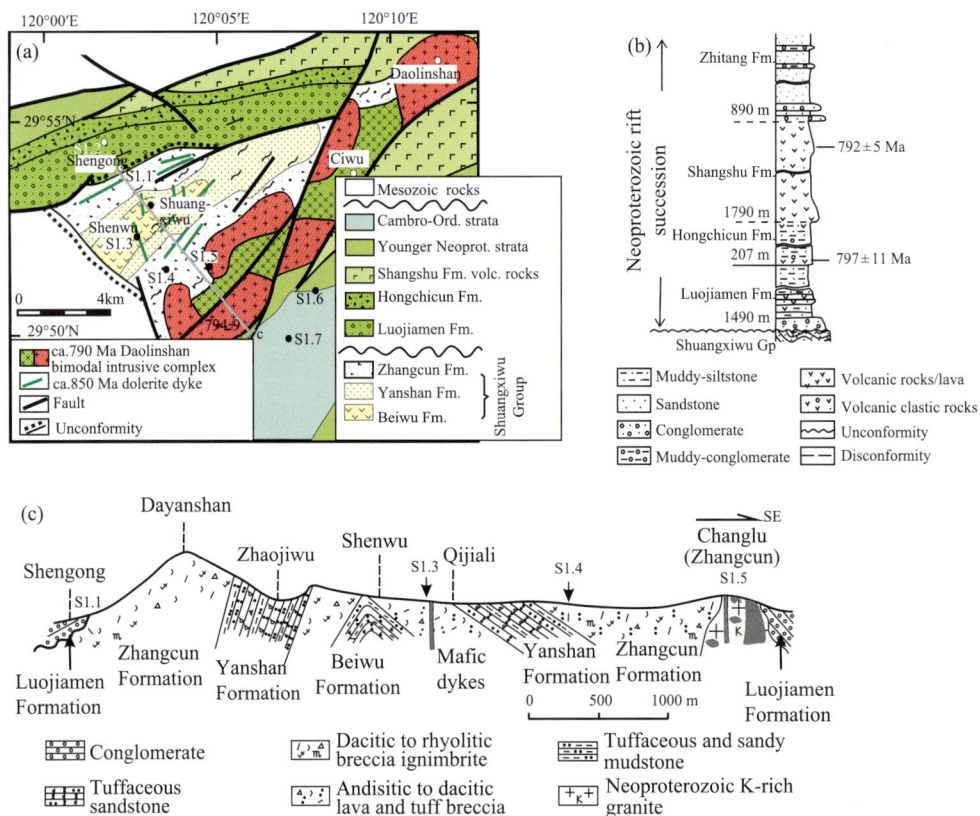

Figure 2.2 (a) A simplified geological map of the Shuangxiwu region (Li et al., 2009), (b) Neoproterozoic stratigraphy of the region (Li et al., 2003a), and (c) a cross-section of the Shuangxiwu anticline (modified from BOGAMR, 1989)

Figure 2.3 Basal conglomerate of the Luojiamen Formation at Shengong village

tions (Figure 2.2(a), (b)) gave overlapping ages of 797 ± 11 Ma and 792 ± 5 Ma, respectively (Li X H et al., 2008; Li et al., 2003a). Consistent ca. 800 Ma ages from various parts of such a thick succession (Figure 2.2(b)) indicate rapid deposition that requires both the availability of accommodation (rift basin) and sufficient supply of materials (strong volcanic activities supplying volcanic and volcaniclastic rocks, and rapid erosion of the rift shoulders).

Stop 1.2   Upper Luojiamen Formation and Cretaceous dike (29°54.138′N, 120°01.631′E)

At this quick stop, we will examine: ① tuffaceous members of the Neoproterozoic Luojiamen Formation with horizontal laminations; and ② a Cretaceous dike intruding the formation. This locality lies in the middle to upper parts of the Luojiamen Formation which is characterized by fine-grained tuffaceous rocks. The volcanic ashes have been altered to chlorite, epidote and siliceous aggregates. The monzonitic granitic porphyry dike intruding the Luojiamen Formation shows clear cooling joints, and has a porphyritic texture of plagioclase, K-feldspar and quartz. SHRIMP zircon U-Pb dating of the dike gives a $^{206}Pb/^{238}U$ weighted mean age of 118 ± 3 Ma (J. Wang and others, unpublished results as quoted in Li et al., 2003b). This indicates that many of the dikes in the region were part of the late Mesozoic magmatic events (see Section 1.7 and Day 3 program).

Stop 1.3   Beiwu Formation arc volcanic rocks and 850 Ma post-orogenic doleritic dikes (29°52.120′N, 120°02.733′E)

The Shuangxiwu Group consists predominantly of volcanic and pyroclastic rocks interbedded with felsic tuff and tuffaceous sandstones and siltstones that were strongly deformed. It has been divided into four formations according to lithologic characteristics, including, from bottom to top, the Pingshui, Beiwu, Yanshan and Zhangcun formations (BOGAMR, 1989). The Pingshui Formation is only exposed around Pingshui. It consists chiefly of altered andesitic rocks (Figure 2.4(a)). It is regarded as the lowest formation in the group, although it has no direct contact with the other three formations. The minimum age of the Pingshui Formation is constrained by a SHRIMP zircon U-Pb age of 913 ± 15 Ma from the Taohong pluton that intrudes the Pingshui basalts (Ye et al., 2007). Chen et al. (2009) reported LA-ICPMS zircon U-Pb analyses of two volcanic rocks from the Pingshui Formation. While the authors interpreted their average $^{206}Pb/^{238}Pb$ ages of 904 ± 8 Ma and 906 ± 10 Ma as the formation age of the Pingshui volcanic rocks, it is noted that their measured U-Pb data are highly discordant, with $^{206}Pb/^{238}Pb$ ages ranging from 878 Ma to 999 Ma. On the other hand, the two dated samples have nearly identical $^{207}Pb/^{206}Pb$ ages within errors, with a mean of 965 ± 12 Ma. This $^{207}Pb/^{206}Pb$ age, consistent with the Sm-Nd internal isochron age of 978 ± 44 Ma (Zhang et al., 1990), should be considered the best estimate of the formation age of the Pingshui volcanic rocks (Li et al., 2009a).

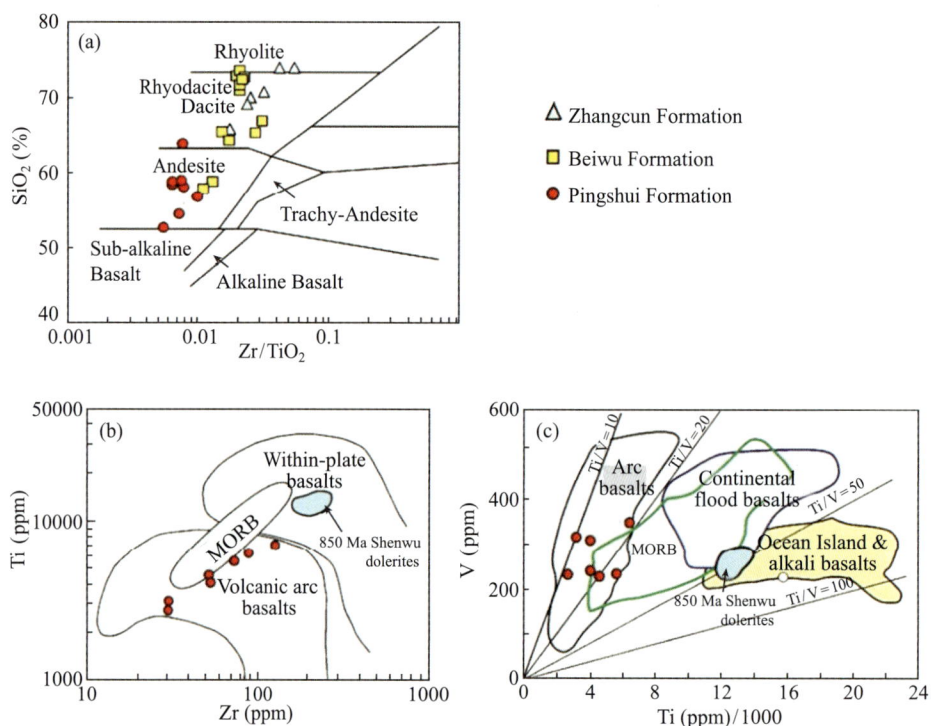

Figure 2.4   (a) Rock classification of the Shuangxiwu Group volcanic rocks; (b) Ti vs. Zr discrimination diagram of Pearce (1982); and (c) V vs. Ti discrimination diagram of Shervais (1982) for the Pingshui basaltic rocks. The fields of arc basalts, MORB, continental flood basalts, and ocean-island and alkali basalts were drawn by Rollinson (1993) according to Shervais (1982) (figure from Li et al., 2009)

Li et al. (2009) illustrated that the Shuangxiwu Group volcanic rocks and associated intrusive tonalites and granodiorites, dated at between ca. 965 ± 12 Ma and 891 ± 12 Ma, constitute a typical calc-alkaline magmatic assemblage of an active continental margin (Figure 2.4).

At this stop, we first examine the Beiwu Formation at the core of the Shuangxiwu anticline, ca. 50 km to the west of Pingshui.   It has a thickness of ca. 430 m, consisting of andesite, dacite, rhyolite and volcaniclastic interbeds. It has a SHRIMP zircon U-Pb age of 926 ± 15 Ma (Li et al., 2009; Figure 2.5(a)).

At this stop, we will also see little-deformed, sub-vertical mafic dikes intruding the Shuangxiwu Group arc volcanic rocks. One of the dikes is dated at 849 ± 7 Ma (Figure 2.5(b)). Geochemical characteristics of the dikes resemble those of intraplate basaltic rocks in continental rifts (Figure 2.4(b), (c), Figure 2.6(a), (b); Li et al., 2009). Therefore, it appears that the Shuangxiwu volcanic arc terminated after ca. 890 Ma, and by ca. 850 Ma the tectonic regime of the region had already been transformed from a compressional one into an extensional one (Li et al., 2009).

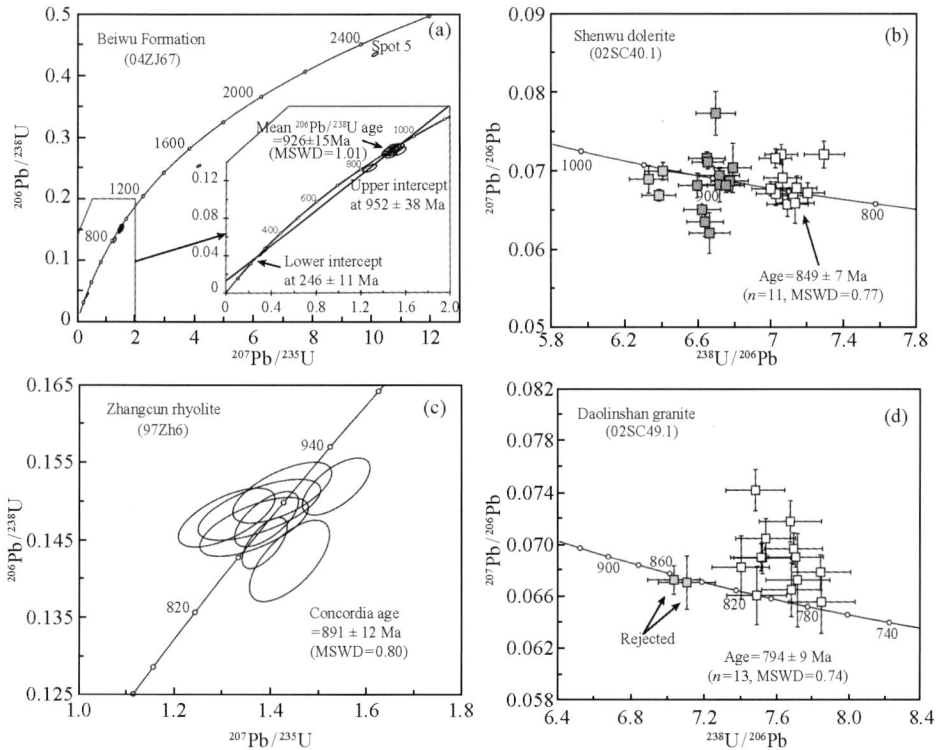

Figure 2.5   SHRIMP U-Pb zircon age of the Beiwu Formation rhyolite sample (a), the Shenwu dolerite dike (b), and the Zhangcun Formation (c), and the Daolinshan granite (d)   (Li et al., 2009)

Stop 1.4   Zhangcun Formation ignimbrites ─ the last of the Shuangxiwu arc magmatism (29°51.834′N, 120°03.789′E)

Zhangcun Formation, the uppermost formation of the Shuangxi Group, consists mainly of felsic ignimbrite (Figure 2.4(a)) with a thickness of ca. 850 m. It has been dated using the SHRIMP zircon U-Pb method at 891 ± 12 Ma (Figure 2.5(c); Li et al., 2009).

The significance of this stop is that the Zhangcun Formation is the youngest, non-controversial, Neoproterozoic arc volcanic rocks in South China. It may thus signify the end of the Shuangxiwu arc when the ocean between the Yangtze and Cathaysia blocks was closed (Figure 1.3, Figure 1.5, Figure 1.7(b)).

Stop 1.5   Ca. 790 Ma Daolinshan granite-diabase complex ─ part of the Neoproterozoic bimodal magmatism (29°51.673′N, 120°04.787′E)

The Daolinshan granite-diabase complex crops out in a northeasterly regional extend of ca. 5×30 km (Figure 2.2). It consists of pinkish K-feldspar granites and dark grey diabases. The complex intrudes the Shuangxiwu Group. In the field, the diabases are mostly in intrusive contact with the granites, whereas granitic veins are also visible within the diabase bodies near the contact zone. The contacts between the diabases and the granites are generally irregular. Evidence of ductile interaction between the diabases and the granites they intrude has been found at several outcrops. All these observations suggest that the mafic magma was coeval, and in places likely comagmatic, with the K-feldspar granites. A granite sample gave a SHRIMP U-Pb zircon age of 794 ± 9 Ma (Figure. 2.2(a), Figure 2.5(d); Li X H et al., 2008).

Li X H et al. (2008) found that the studied mafic rocks are all tholeiitic in compositions. They were likely originated from a common asthenospheric mantle source with the Shangshu basalts in the region. Assimilation of different crustal components played a minor role in the genesis of these basaltic rocks. These authors also found that the Daolinshan K-feldspar granites and the Shangshu rhyolites show close affinities to aluminous A-type granites. They were generated by shallow dehydration melting of the preexisting Sibaoan arc calcalkaline igneous rocks with variable degrees of fractional crystallization. The igneous rocks are anorogenic in origin (Figure 2.6). Li X H et al. (2008) further suggested that the ca. 850 Ma Shenwu dolerites may represent the initiation of the anorogenic magmatism that predated the large-scale onset of the Nanhua rifting, whereas the ca. 790 Ma bimodal magmatism occurred during the middle of the rifting. Wang et al. (2010) reported 780 Ma A-type granites in the eastern part of the Daolinshan igneous complex, and interpreted them as the products of high-temperature melting of slightly older (850–800 Ma?) tholeiitic rocks underplated in the lower crust, all related to plume events during the breakup of the supercontinent Rodinia (Li Z X et al., 2008).

Optional stops: On the way to Zhuji in the late afternoon, if time allows, we may make ad-hock (Stops 1.6 and 1.7) along the way to examine mid-Neoproterozoic to early Paleozoic strata on the roadsides. Note that the two localities are between the Shuanxiwu arc and the Jiang-Shao Fault (see Day 2 program).

Stop 1.6 Neoproterozoic upper rift succession (Zhitang Formation volcaniclastic rocks) in fault contact with basal Cambrian (Hetang Formation) carbonaceous shale (29°50.968′N, 120°7.811′E). There is no precise age for the formation here, but lateral correlations suggest a ca. 750 Ma age.

Stop 1.7 Middle Cambrian Yangliugang Formation: moderately deformed but non-metamorphosed, thin- to medium-bedded limestone (29°49.832′N, 120°6.988′E).

Exercise of the day:

- Draw a simple tectonostratigraphic column for the Shuangxiwu region, using absolute ages to define the vertical axis.

Question of the day (for thinking about during the day, and for discussion in the evening over a drink):

• How to distinguish subduction-related magmatism from rift or plume magmatism using geological and geochemical information?

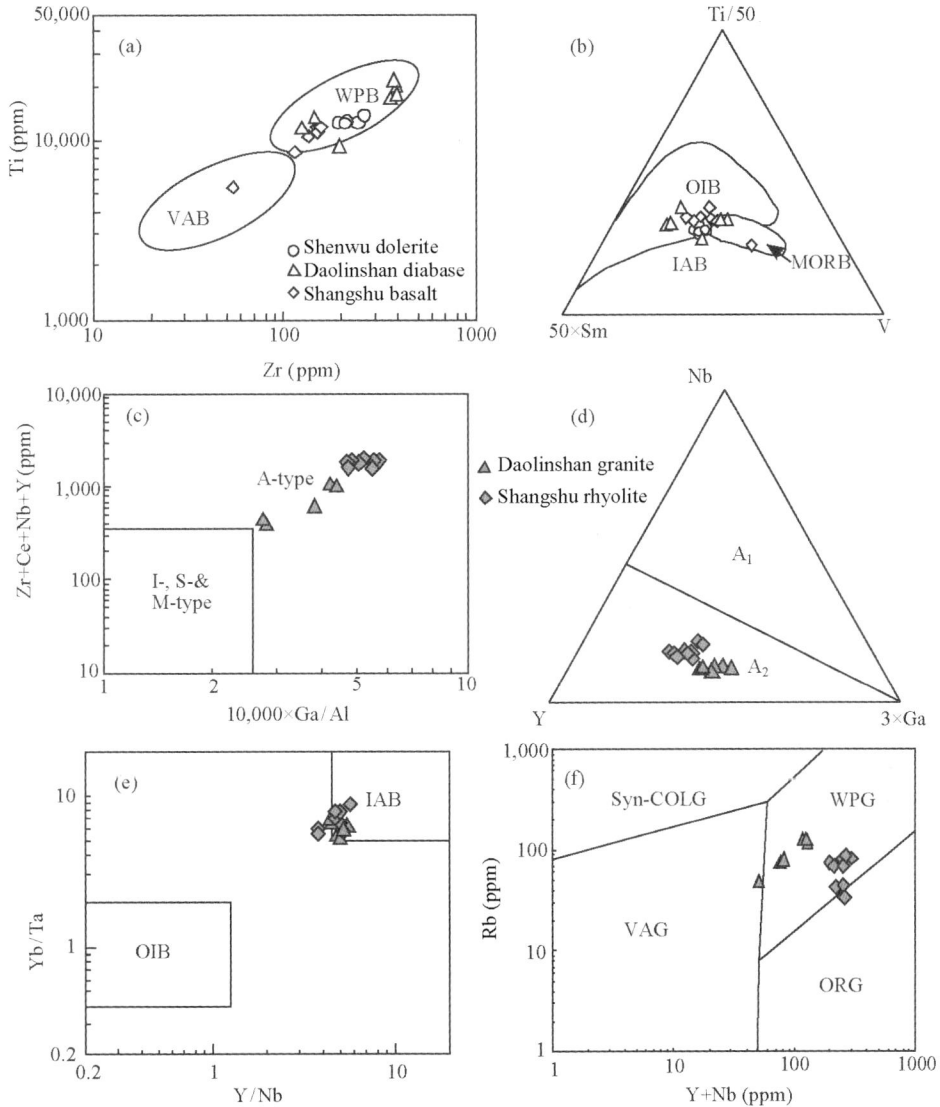

Figure 2.6   Geochemical discrimination diagrams of (a) Ti-Zr (Pearce,1996) and (b) Ti-Sm-V (Vermeesch, 2006) for the northern Zhejiang mid-Neoproterozoic mafic igneous rocks. (c)–(f) are plot of the northern Zhejiang mid-Neoproterozoic felsic igneous rocks: (c) (Zr+Ce+Nb+Y) vs. 10,000×Ga/Al diagram (modified after Whalen et al., 1987) showing an A-type granite affinity; (d) Nb-Y-Ga ternary diagram (Eby, 1992) where all data fall into the A2-subtype granite field; (e) Yb/Ta vs. Y/Nb diagram (Eby, 1992) exhibiting similaritifes to the IAB; and (f) Rb vs. (Y+Nb) diagram (Pearce et al., 1984) where all data plot into the within plate field (Figures are from Li X H et al., 2008)

➤Day 2. Chencai Complex: basement rocks of Cathaysia, Early Paleozoic Wuyi-Yunkai Orogeny, and the Jiangshao Fault

Aim: We will examine the protolith of rocks regarded as representing the Cathaysia Block, the timing and nature of the early Paleozoic Wuyi-Yunkai ("Caledonian") Orogeny, and the significance of the Jiangshao Fault – a perceived boundary between the Yangtze and Cathaysia blocks.

Contents: Chencai Complex garnet-rich paragneiss, granitic gneiss (and gneissic granite?), marble, and 1.78 Ga meta-gabbroic dikes/sills and country gneisses (the oldest rocks found in the region); deformation styles of the Chencai Complex; the 840 Ma Lipu/Huangshan gabbro-diorite; the Jiangshao Fault.

The Chencai Complex (traditionally called the "Chencai Group") in central Zhejiang Province consists of gneiss, amphibolite, garnet-muscovite schist, and marble, with a predominantly northwesterly structural vergence (Figures 2.8(a),(b)). The complex has long been an enigma in tectonic studies of the South China Block. Due to its medium- to high-grade metamorphism, its complex lithology and structure, and its uncertain age, it has variably been interpreted as: ① the >900 Ma basement of the Cathaysia Block (Shui, 1987; Zhou and Zhu, 1993), ② a 1400–900 Ma ophiolitic mélange (Li, 1993), or ③ a Meso- to Neoproterozoic arc (Kong et al., 1995). Kong et al. (1995) documented the presence of a large volume of mafic to felsic metavolcanic and intrusive rocks in the complex. The mafic rocks range from tholeiitic basalt to basaltic andesite and tuff. The complex is bounded to the north by the Jiang-Shao Fault (zone) (Figure 2.8), and has no direct contact with known Paleozoic sedimentary rocks. It is unconformably overlain by Early Jurassic and younger successions (Figure 2.7; Zhejiang and BOGAMR, 1989).

Recent work by Li et al. (2010) indicates that the protolith of the complex consists predominantly of Neoproterozoic volcanic and volcaniclastic rocks (Stop 2.7), with little-metamorphosed bimodal volcanic rocks exposed less than 10 km north of Stop 2.7 dated at 838 ± 5 Ma. A sliver of Paleoproterozoic basement intruded by 1781 ± 21 Ma gabbroic sills/dykes had also been identified within the complex (Stop 2.3). Much of the complex suffered early Paleozoic (Wuyi-Yunkai) deformation and metamorphism (Figure 2.8(b)), with amphibolite-phases metamorphism in the region occurred at ca. 450 Ma (stops 2.3 and 2.7), decompressional melting at ca. 430 Ma (Stops 2.3 and 2.5), and hornblende cooling at ca. 425 Ma (see Figure 1.3, Figures 1.16–1.21 and Section 1.6). The protolith age of marbles in the complex (Stop 2.6) is unknown, but could be either late Neoproterozoic or earliest Palaeozoic.

The pluton is cut by several faults (part of the Jiang-Shao fault zone?), and variably deformed. It is considered to be part of the mid-Neoproterozoic bimodal magmatism related to continental rifting (Section 1.5 and Day-1's Stop 1.5), but detailed geochemical and petrographical analyses are yet to be completed (W X Li and others, work in progress).

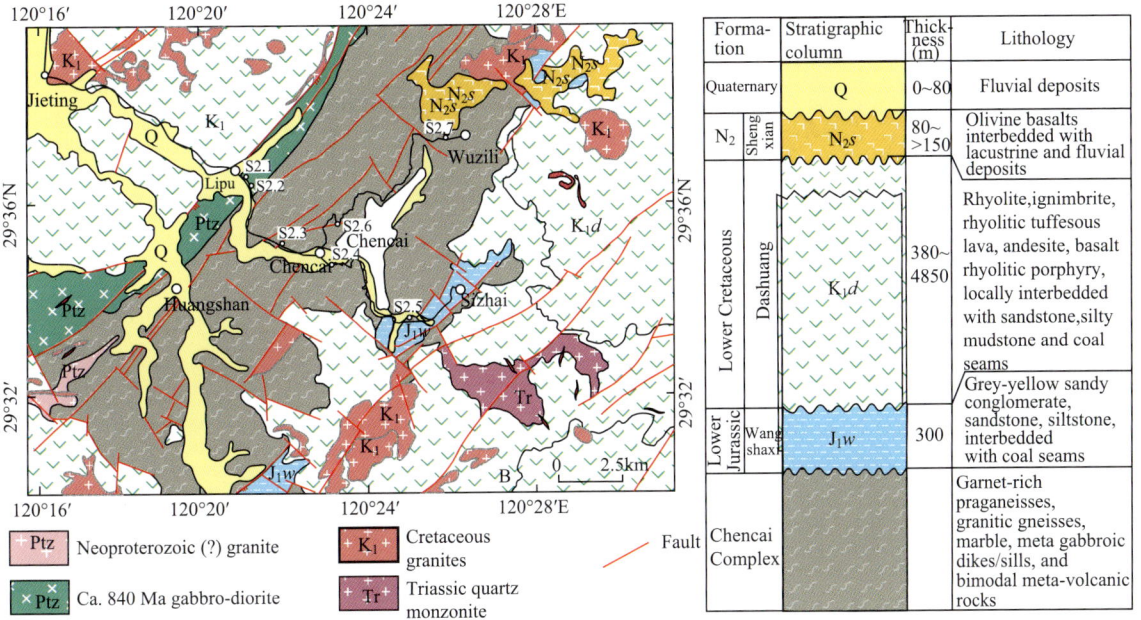

Formation		Stratigraphic column	Thickness (m)	Lithology
Quaternary		Q	0~80	Fluvial deposits
N₂	Sheng xian	N₂s	80~ >150	Olivine basalts interbedded with lacustrine and fluvial deposits
Lower Cretaceous	Dashuang	K₁d	380~ 4850	Rhyolite, ignimbrite, rhyolitic tuffesous lava, andesite, basalt rhyolitic porphyry, locally interbedded with sandstone, silty mudstone and coal seams
Lower Jurassic	Wang shaxi	J₁w	300	Grey-yellow sandy conglomerate, sandstone, siltstone, interbedded with coal seams
Chencai Complex				Garnet-rich praganeisses, granitic gneisses, marble, meta gabbroic dikes/sills, and bimodal meta-volcanic rocks

Legend:
- Ptz — Neoproterozoic (?) granite
- ✕ Ptz — Ca. 840 Ma gabbro-diorite
- K₁ — Cretaceous granites
- Tr — Triassic quartz monzonite
- Fault

Figure 2.7   Geological map of the Chencai region (modified from 1:250,000 Zhuji Geological Map)

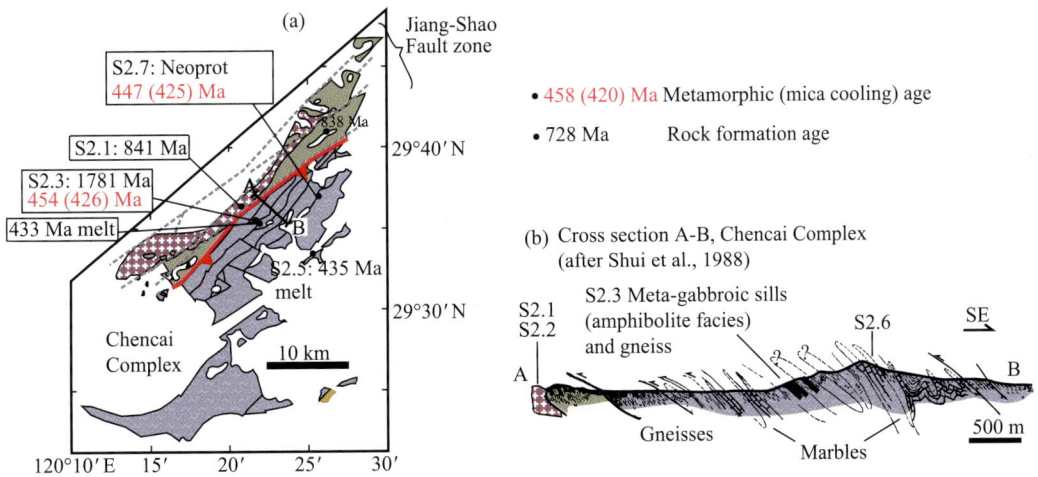

Figure 2.8   Ages (a) and structural styles (b) of the Chencai Complex (Li et al., 2010)

Stop 2.1 Little-deformed 840 Ma Lipu/Huangshan gabbro-diorite (29°36.627′N, 120°20.973′E)

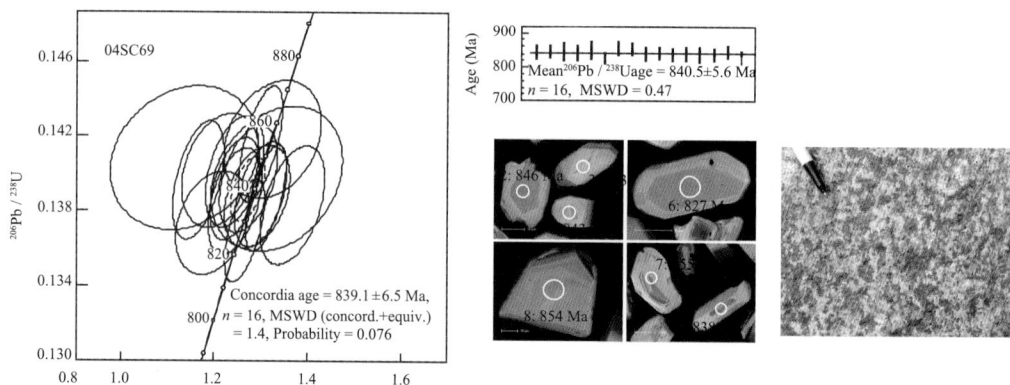

Figure 2.9 SHRIMP U-Pb zircon age of the Lipu/Huangshan gabbro-diorite (S2.1 in Figure 2.8). (Li et al., 2010)

Stop 2.2 A shear zone in the Lipu/Huangshan gabbro-diorite (29°36.377′N, 120°21.244′E)

We observe here a NE-trending, sub-vertical mylonitic shear zone inside the Lipu/Huangshan gabbro-diorite. The shear zone, part of the Jiang-Shao fault zone, has not been dated yet.

Stop 2.3 Paleoproterozoic protoliths in the Chencai Complex – the oldest rocks in this region (29°35.195′N, 120°21.882′E)

Here we observe Paleoproterozoic (1781 ± 21 Ma) gabbroic intrusions (dikes and/or sills) metamorphosed into amphibolites. The 1781 ± 21 Ma upper intercept age is interpreted as the protolith age, and the lower intercept age of 454 ± 29 Ma the high-grade metamorphic event (Figure 1.19, Figure 2.10), which is consistent with the internal zoning patterns of the zircon crystals and the variations in the Th/U ratios (Figure 2.10(a); Li et al., 2010). A migmatitic sample from the nearby country rocks revealed 2700–2200 Ma zircon core ages and a 433 ± 3 Ma migmatization age. Metamorphic hornblendes gave a 426 ± 1 Ma cooling age (Figure 2.10(b)). See Figure 1.19 for the *P-T* condition and *P-T-t* path of the region.

Take a measurement of the dominant tectonic foliation here.

Stop 2.4 Chencai Complex metamorphic rocks or Cretaceous granite? (29°34.736′N, 120°23.525′E)

This is a quick stop to verify whether the rocks exposed along the road beneath the dam are part of the Chencai metamorphic complex, or a Cretaceous granite as shown in a recent geological map. Note the presence of garnets and metamorphosed mafic dikes.

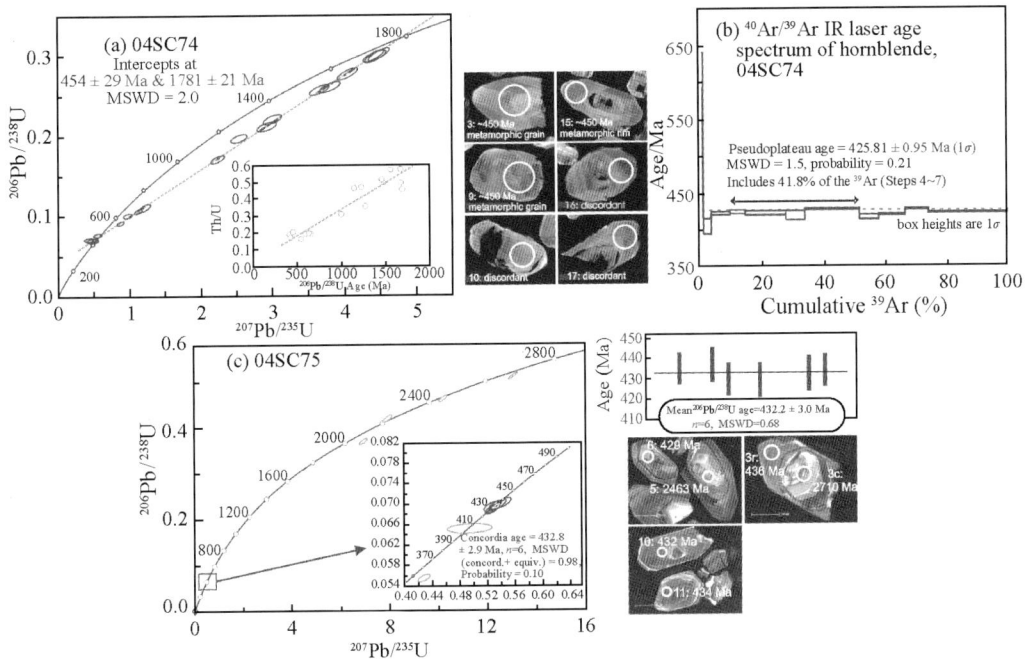

Figure 2.10   SHRIMP U-Pb zircon ages (a) and $^{40}Ar/^{39}Ar$ IR laser age spectrum of hornblende (b) of a meta-gabbroic dike in the Chencai Complex, and ages from a nearby migmatite sample (c) (Li et al., 2010)

Take a measurement of the dominant tectonic foliation here and compare it with that at Stop 2.3.

Stop 2.5   Chencai Complex granitic gneiss/gneissic granite (29°33.739′N, 120°25.163′E)

Under the bridge leading to Sizhai, there is a good outcrop of granitic gneiss and/or gneissic granite. The granitic gneiss is garnet-rich and appears to be a paragneiss (the top-left picture in Figure 2.11). Toward the downstream of the creek across the bridge, the rocks have more of a gneissic granite texture (the top-right picture in Figure 2.11). Zircon grains from the gneissic granite have only a few ca. 800 Ma and older cores, with magmatic-like zircon zoning patterns (Figure 2.11) and moderate Th/U ratios. The 435 ± 4 Ma age (Figure 2.11) was interpreted to be the age of syn- to late-orogenic melting. Note that this age is consistent with the 433 ± 3 Ma migmatization age near Stop 2.3 (Figure 2.10(c)).

In a recent geological map the site is located at the edge of a Cretaceous granite – a suspected misinterpretation.

Take a measurement of the dominant tectonic foliation here and compare it with those in the previous two stops.

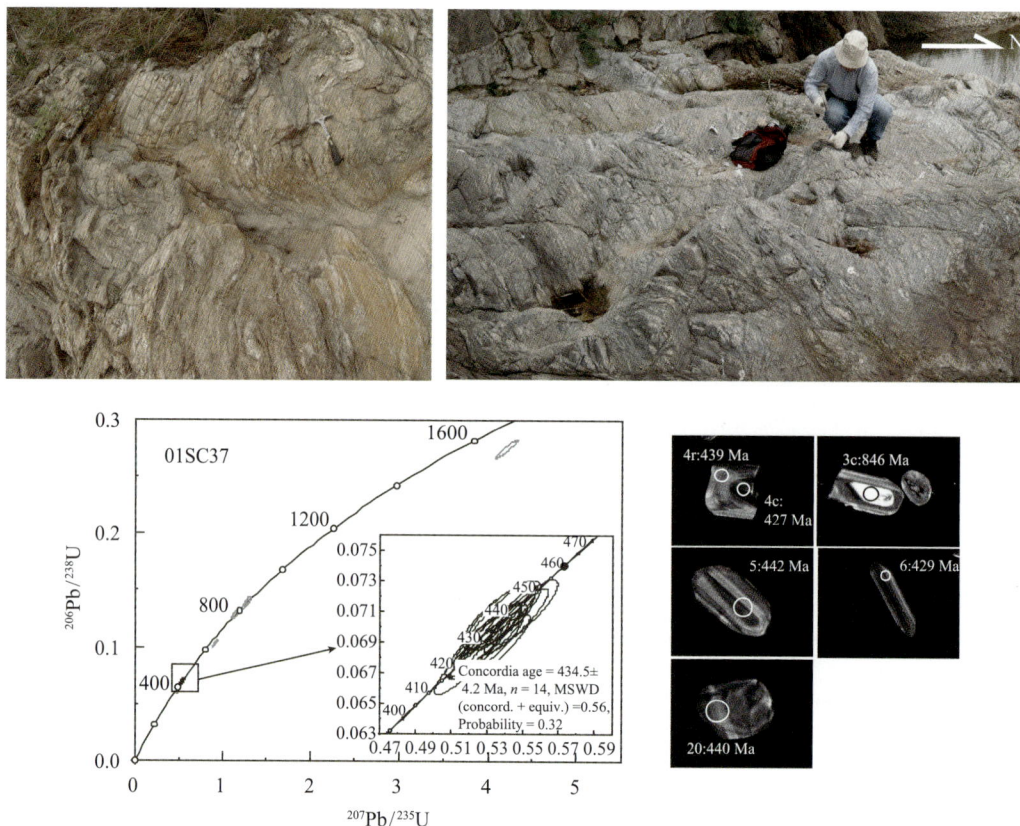

Figure 2.11　Field photos and SHRIMP U-Pb zircon ages of a gneissic granite sample near Sizhai (Li et al., 2010)

Stop 2.6　Chencai Complex marble outcrop (29°35.431′N, 120°23.088′E)

Here we examine along the road cut marble outcrops as part of the Chencai Complex. Amphibolite are found amongst marble in a deserted marble quarry further east at the sharp turn of the road (place for a comfort stop). The rocks have not been dated, but the protolith could be comparable to the Neoproterozoic Mamianshan "Group" in northwestern Fujian, which contains marbles, schists and metavolcanic rocks dated at 818 ± 9 Ma (Li et al., 2005). However, the protolith for the marble could be significantly younger too.

Stop 2.7　Chencai Complex garnet-rich paragneiss at Wuzili village (29°37.360′N, 120°26.002′E)

In the creek bed beneath the stone bridge we see garnet-rich paragneiss (Figure 2.12, sample 01SC40), featuring dominantly Neoproterozoic detrital ziron ages and a 447 ± 7 Ma amphibolite-phases metamorphic age. $^{40}Ar/^{39}Ar$ UV laser spot dating of biotite grains from sample 04SC88, a garnet-rich gneiss sample that is similar to and was collected close to sample

01SC40, showed a systematic variation in ages from core to rim, with a maximum core age of 425 ± 4 Ma (Li et al., 2010). The 425 ± 4 Ma biotite core age agrees within error with the higher temperature (ca. 500 °C) hornblende cooling age of 426 ± 1 Ma from sample 04SC74 of the Chencai Complex, which suggests that cooling from 500 °C to 300 °C was very rapid (Li et al., 2010).

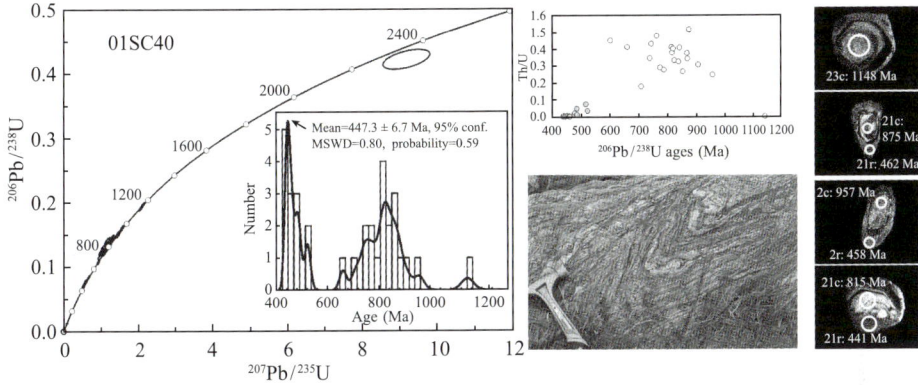

Figure 2.12   SHRIMP U-Pb dating of the garnet-rich Chencai paragneiss at Wuzili (Li et al., 2010)

Exercise of the day: Draw a simple tectonostratigraphic diagram for the Chencai region, and compare it with that of the Shuangxiwu region (use the same style and vertical scale for all columns).

Questions of the day:

- Where does the boundary between the Yangtze and Cathaysia blocks lie? What basement you would expect for the region between the Shuangxiwu arc and the Jiang-Shao Fault: Yangtze, Cathaysia or either?
- Why there do not appear to be any Paleozoic rocks south of the Jiang-Shao Fault?

Snow over the Huangdu (Emperor Crossing) Bridge (S3.3) (Zheng-Xiang Li)

➤Day 3. Late Mesozoic (post P-Tr Indosinian Orogeny) large magmatic province

Aim: We will examine the Cretaceous bimodal volcanic and intrusive rocks as part of the late Mesozoic large magmatic province in eastern South China, and discuss their tectonic significance.

Contents: The Jingling Cretaceous bimodal volcanic succession (with exciting volcanic structures), the Ru'ao bimodal intrusive complex, and Cretaceous tree fossils and host rock.

Jurassic–Cretaceous magmatism was widespread in central and eastern South China, but their tectonic significance is controversial (see Section 1.7 and Figure 1.25). Although Mesozoic plutonic rocks in eastern Zhejiang goes much further back in time, the volcanic rocks, blanketing much of the landscape, are dominantly of Cretaceous ages. The region constitutes a world-class paleo-volcanic field. and there is a number of geoparks setup to protect some of the best volcanic structures and landscape. However, to our best knowledge, there is yet not a single book dedicated to documenting the fantastic volcanic structures and paleogeographic history of the region.

Stop 3.1 Early Cretaceous bimodal volcanic succession at Jingling, Xinchang County (29°21.785′N, 120°46.642′E)

We will start from the base of the hill south of Jingling town (Figure 2.13) and walk toward the gap at the top of the ridge, examining the Early Cretaceous Guantou Formation bimodal volcanic rocks and volcanic structures (Figure 2.14).

We will examine two layers of basaltic flows with lava tubes, agglomerate, and ignimbrite (welded tuff) with rhyolitic textures. We will see agates in a massive tuff layer at the pass, a result of silicate concentration by fluids percolating through the volcanic ashes.

Stop 3.2 Xinchang fossil woods geopark, Chengtan, Xinchang (29°23.713′N, 120°46.850′E)

We will examine some giant petrified tree trunks in the Early Cretaceous strata (Figure 2.15(a)) and possible causes for the events. Six layers of petrified woods have been discovered on this cross section, all within the middle section of the Guantou Formation. Some of the giant fossilized tree trunks, with or without roots remaining, lie sub-horizontally parallel to the bedding. Others have the stump only left, but still standing.

Tuff samples from the third petrified layer have been dated at ca. 119~116 Ma using SHRIMP U-Pb zircon method (Zhang F Q et al., unpublished results).

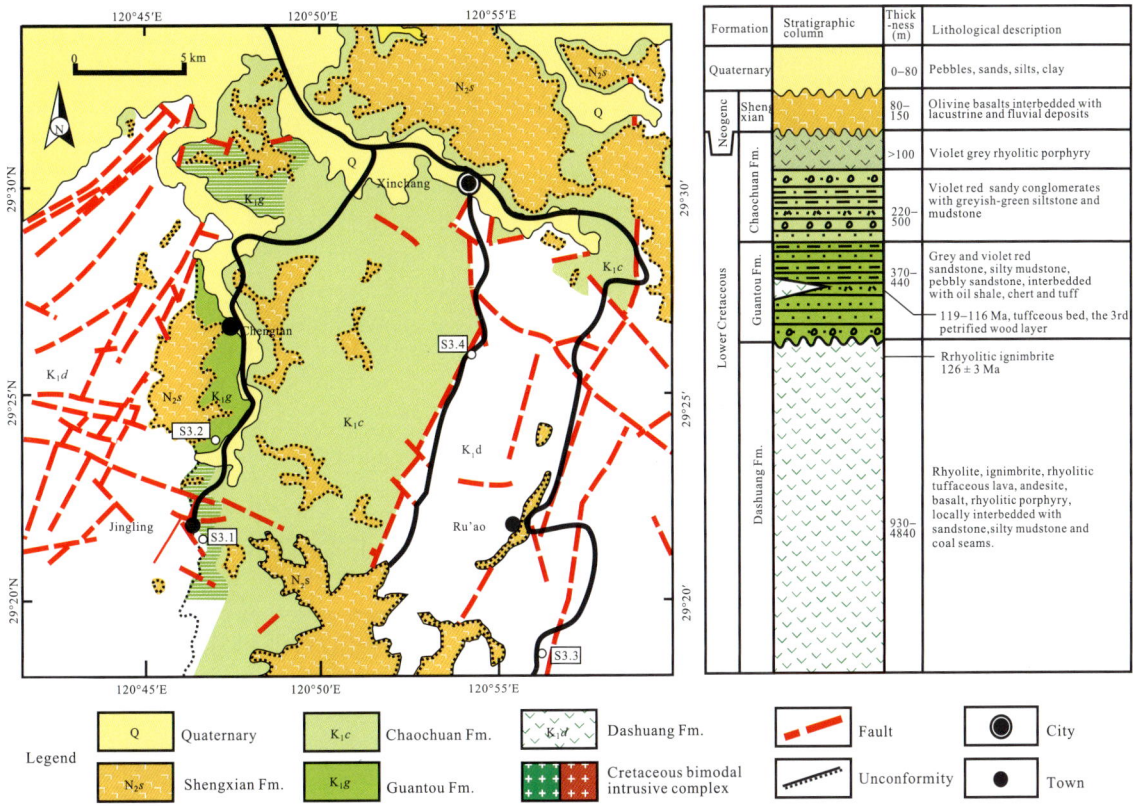

Formation		Stratigraphic column	Thick-ness (m)	Lithological description
Quaternary			0–80	Pebbles, sands, silts, clay
Neogene	Sheng-xian		80–150	Olivine basalts interbedded with lacustrine and fluvial deposits
Lower Cretaceous	Chaochuan Fm.		>100	Violet grey rhyolitic porphyry
			220–500	Violet red sandy conglomerates with greyish-green siltstone and mudstone
	Guantou Fm.		370–440	Grey and violet red sandstone, silty mudstone, pebbly sandstone, interbedded with oil shale, chert and tuff
				— 119–116 Ma, tuffceous bed, the 3rd petrified wood layer
	Dashuang Fm.			Rrhyolitic ignimbrite 126 ± 3 Ma
			930–4840	Rhyolite, ignimbrite, rhyolitic tuffaceous lava, andesite, basalt, rhyolitic porphyry, locally interbedded with sandstone, silty mudstone and coal seams.

Figure 2.13   Geological Map of the Xinchang region (modified after 1:250,000 Shengxian and 1:200,000 Zhuji geological maps)

A previous theory about the formation of the petrified woods was that volcanic activities in the region drove fluids through old quartzite, and the siliceous fluids petrified the giant trees. However, it appears that the petrified woods are commonly set in volcanic tuff layers. Li Z X et al. (this work) thus speculate that hot volcanic ashes, and possibly heated water both above and below ground, may have killed the trees, and the silica leached out of the ashes (as shown by the large amount of agates in the tuffs and ignimbrite at Stop 3.1) petrified the trunks.

The presence of petrified trunks in the region indicates that at least the topographic lower parts of the region were below the tree line (presently at ca. 1500 m) – an important paleo-elevation indicator.

Figure 2.14 (a) A basaltic lava flow with dense basalt in the core surrounded by an amygdaloidal basaltic shell. (b) An interpreted lava tube filled with volcanic breccia (d) and possibly some younger flows (c) (Li Z X et al., this work)

Figure 2.15   (a) A petrified tree trunk in the Chengtan geopark. (b), (c) A present day example of volcanic ashes erupted from the Hudson volcano in 1991, killing a large track of forest along the Ibañez River valley ca. 50 km away from the volcano in southern Andes (Li Z X, 2006 photos). (d) Stratigraphic location of the fossil woods layers (Zhang F Q et al. unpublished work)

Stop 3.3 Early Cretaceous Ru'ao bimodal intrusive complex (29°18.750′N, 120°56.367′E)

Here we see evidence of magma mingling between granite and dolerite (Figure 2.16). In the riverbed to the west we see the dolerite intrusive body, and to the east is the Ruao granite (see good outcrop at the end of the bridge), dated at 116 ± 3 Ma using the SHRIMP U-Pb zircon method (Dong et al., 2008).

If time allows, we could drive into the central part of the granite to see mafic dikes intruding the granite.

Both the Cretaceous volcanic rocks and magmatic intrusions in the broad region feature bimodal compositions. Although andesitic rocks are present, they only account to a small proportion of the late Mesozoic igneous rocks. This raises the question of what tectonic environments these igneous rocks were formed in. Popular models include Andean-type continental arc (Jahn et al., 1990; Zhou and Li, 2000; why the bimodal magmatism then?), basin-and-range style extension (Gilder et al., 1991), or a combination of basin-and-range style extension due to the foundering of a previously subducted flat-slab, and a reinitiated new continental arc (Li and Li, 2007; Li et al., 2012; see Section 1.7).

Figure 2.16   Magma mingling textures in the Ru'ao intrusive complex, Xinchang

Stop 3.4 Cooling joints in the Early Cretaceous basalt lavas near Bali village, southern Xinchang (29°25.937′N, 120°54.328′E)

Here we see well developed columnar cooling joints of basalts at the bottom section of the Early Cretaceous Dashuang Formation (Figure 2.13, Figure 2.17). Such jointing is attributed to post-solidification thermal contraction of hot lava. Polygonal joints typically develop perpendicular to cooling isotherms (such as bedding planes of lava flows, roof or floor planes of sills, or wall planes of dikes – see Stop 1.2) to accommodate the contraction.

Figure 2.17   Cooling joints in the Early Cretaceous basalt lavas near Bali village, southern Xinchang (Zhang F Q)

Exercise of the day: Draw a cartoon diagram illustrating the paleoenvironment of the region based on what you have seen today and the regional geological map.

Question of the day:

- Could you think of a place in the present world where the tectonic environment and landscape may resemble the coastal region of eastern South China in late Mesozoic?

Local Danxia landform in Cretaceous strata (Zheng-Xiang Li)

➤Day 4. Early Paleozoic strata in the Jiangshan region: Onset of the Wuyi-Yunkai Orogeny and structures of the Indosinian Orogeny

Aim: Early Paleozoic rocks and how they, along with the angular unconformity between Odovician carbonates and Carboniferous basal conglomerates, have recorded the onset and the actual impact of the Wuyi-Yunkai Orogeny in the region.

Contents: Cambrian carbonates, Ordovician muddy limestone, mudstones/shale, and angular unconformity with Lower Carboniferous arkosic sandstone, along the N-S Jiangshan-Dachen road. The traverse cuts across a number of NE-trending folds.

Drive from Xinchang to Jiangshan in the morning.

This region is north of the Jiangshan-Shaoxing (or Jiang-Shao) Fault, featuring successions of late-Precambrian to Paleozoic strata with a dominantly NE-trending structures (Figure 2.18(a),(b)). Xiao and He (2005) carried out a systematic structural analysis of the region, and interpreted the widespread NW-verging thrusts in the region as reflecting the Mesozoic suturing between the Yangtze Block (or a Lower Yangtze sub-block; see figure 17 of Xiao and He, (2005)) and the Cathaysia Block, thus supporting the Hsü et al. (1988) model. However, the presence of an angular unconformity between Early Paleozoic and Carboniferous strata in the Jiangshan region (see Figure 2.18(a) and Stop 4.4) suggests that at least in the Jiangshan region (Southeast and Central structural zones of Xiao and He (2005); see Figure 2.18(a),(b)) there was an episode of early Paleozoic (Wuyi-Yunkai) deformation in the pre-Carboniferous strata (Figure 1.16, Figure 1.21). Li and Li (2007) interpreted the Mesozoic thrusting as due to flat-slab subduction (Figure 1.25, Figure 1.25).

Stops 4.1 to 4.4 are within the SE Zone defined by Xiao and He (2005), whereas Stops 4.5 and 5.1 are within their Central Zone.

Stop 4.1 Lower Cambrian Dachenling Formation massive dolomitic limestone (28°49.028′N, 118°35.708′E)

The outcrop is on the eastern side of the road just south of Dachen-xiang. The massive dolomitic limestone dips steeply to the north (Figure 2.19).

Stop 4.2 Middle Cambrian Yangliugang Formation grey muddy limestone, thick dolomites, and banded limestone (28°48.627′N, 118°35.722′E)

Mid-Cambrian strata here are still dominated by platform carbonates. Observe the lithology (Figure 2.20) and take some bedding measurements.

Stop 4.3   Lower Ordovician calcareous nodular mudstones and mudstones (28°48.1.7′N, 118°36.112′E)

Here, for the first time, the sedimentary facies changed from platform carbonates to clastic rocks, interpreted as indicating the change of sedimentary environment from a carbonate platform to a distal foreland basin.

We start this stop from the lower part of Lower Ordovician strata, featuring calcareous nodular mudstones (Figure 2.21(a)–(c)). Walk along the road further up in the Lower Ordovician succession, we will see folded and cleaved nodular mudstones (axial planar cleavage?), and folded mudstones (Figure 2.21(d)). Take some bedding and cleavage measurements as you go to enable you to work out the regional structural trends, analyze structural generations, and work out the tectonic position of the study region in relation to an orogen.

Stop 4.4   Angular unconformity between Middle Ordovician and Early Carboniferous strata (28°43.93′N, 118°35.498′E)

The angular unconformity between the Early Carboniferous Yejiatang Formation arkosic sandstones and mid-Ordovician here signifies the impact of the Wuyi-Yunkai Orogeny (see Section 1.6) in this region. Note the difference in the bedding attitudes on each side of the unconformity (Figure 2.22).

Take some bedding measurements in the strata above and below the unconformity if possible.

This outcrop illustrates that the structures in the Early Paleozoic strata along this traverse (Stops 4.1 to 4.4, and 5.1) are likely dominated by Wuyi-Yunkai (Ordovician-Silurian) deformation, with only a moderate overprint by the Permo-Triassic Indosinian Orogeny (Section 1.7) as shown by the rather shallow bedding attitudes for the Carboniferous strata here.

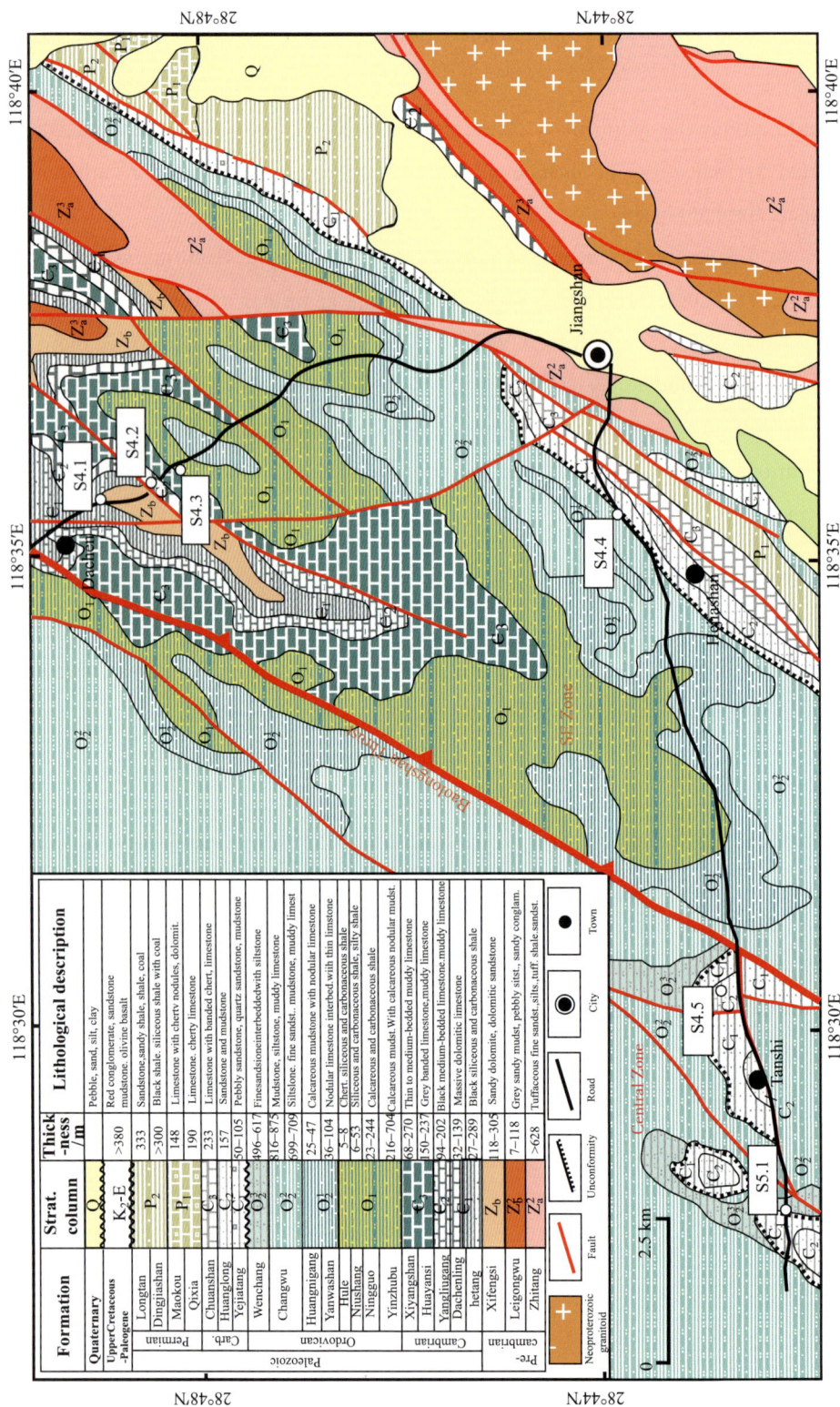

Figure 2.18(a)   Geological map of the Jiangshan region (after 1:200,000 geological map), showing field stops for Day 4 and the first stop for Day 5. The definition of structural zones (i.e., the Central Zone and the SE Zone) follows that of Xiao and He (2005). Note the presence of pre-Carboniferous (Wuyi-Yunkai event) in both Central and SE zones, which is in sharp contrast with the deformation history in the NW Zone (Figure 2.24)

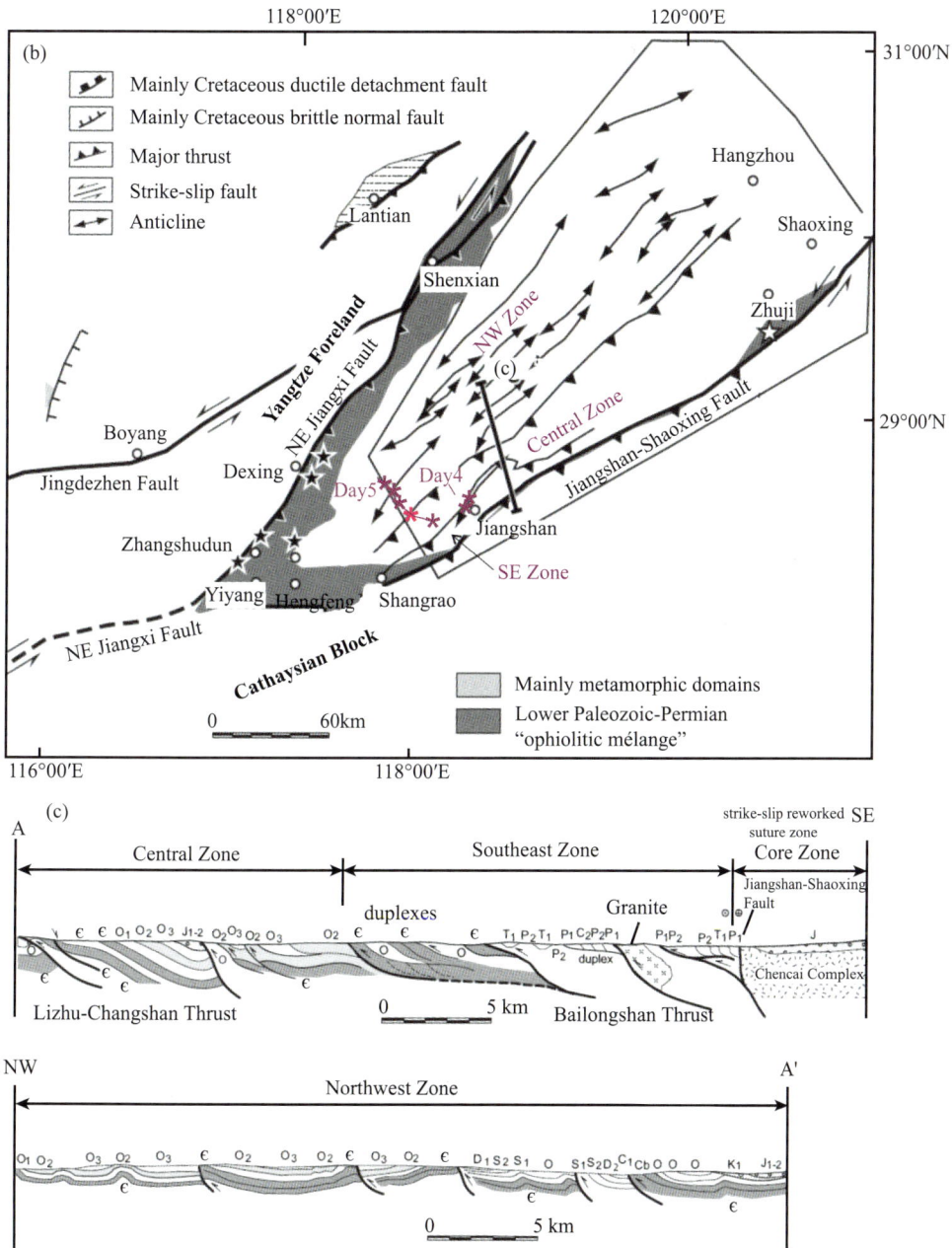

Figure 2.18   (b)(c) Structural map and cross section showing structural styles and tectonic zoning of NW Zhejiang foreland fold-and-thrust belt. A and A′ have approximately the same SE-NW cross-sectional position. K_1–Lower Cretaceous; J_{1-2}–Lower–Middle Jurassic; J–Jurassic; T_1–Lower Triassic; P_1–Lower Permian; P_2–Upper Permian; C_2–Middle Carboniferous; D_2–Middle Devonian; D_1–Lower Devonian; S_1–Lower Silurian; S_2–Upper Silurian; O_3–Upper Ordovician; O_2–Middle Ordovician; O_1–Lower Ordovician; ∈–Cambrian. Straws represent Precambrian basement (modified after Xiao and He, 2005)

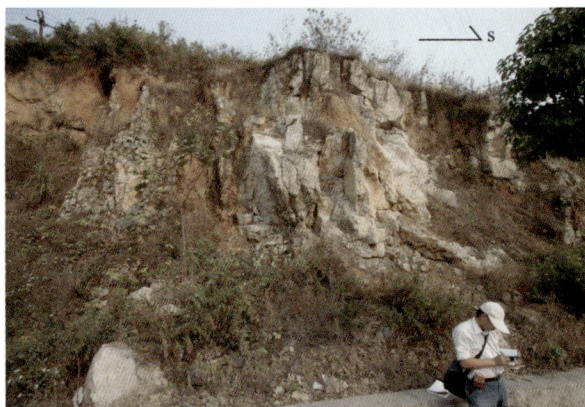

Figure 2.19   Early Cambrian massive dolomitic limestone

Figure 2.20   Yangliugang Formation dolomite (left) and thin bedded limestone (right)

Figure 2.21   Early Ordovician calcareous nodular mudstones (a)–(c) and mudstone (d)

Now have a look of Figure 2.18(a). you will notice that although the structural trends shown by the Late Paleozoic strata in this region are sub-parallel to that of pre-Carboniferous strata, the folds of the older (pre-Carboniferous) strata are much tighter than the younger strata (best shown at the lower left-hand corner of Figure 2.18(a)). We will see a similar relationship at Stop 5.1. This is consistent with the Indosinian deformation in this region being mainly of thin-skinned style (see Figure 2.18(c)), and thus less intense/pervasive in comparison to the Wuyi-Yunkai deformation.

Another interesting observation to be made here is of "mud cracks" and loading structures in the upper part of the mid-Ordovician succession (Figure 2.22(a), (b)), where the lithology changes rapidly from massive carbonate to mudstone.

Figure 2.22   Wuyi-Yunkai unconformity west of Pengli

Stop 4.5   Indosinian deformation in a mid-Carboniferous unit (28°43.041′N, 118°30.478′E)

Between Hengdu and Tanshi, there is a large limestone quarry north of the dirt road, exposing a nice thrust on one side of a box fold in the mid-Carboniferous Huanglong Formation (Figure 2.23; the full geometry of the box fold is not shown), reflecting "Indosinian" deformation here (Should it be called the Huanan deformation/orogeny? See the ca. 1300 km South China fold belt/orogen as in figure 1 of Li and Li, 2007, and major thrusts shown in Figure 1.25(a)). If it is too late to see this in Day 4, we may visit this locality first thing in the following morning.

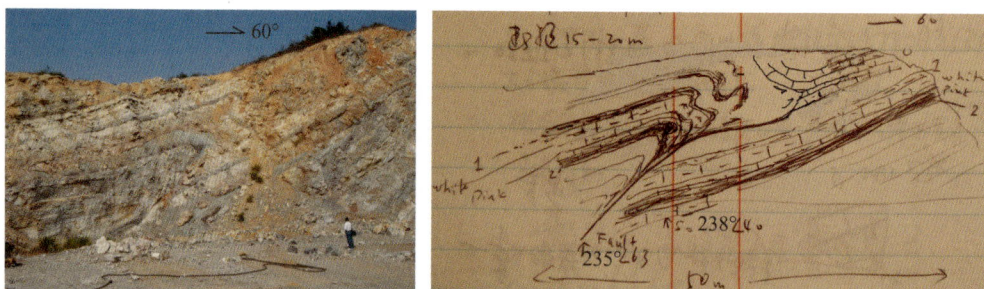

Figure 2.23   An "Indosinian" thrust in mid-Carboniferous rocks near Tanshi, NW of Jiangshan

Exercises of the day:

- Draw a simple tectonostratigraphic column for this region, following the same style and vertical scale as in Day 1 and Day 2.
- Work out the fold axis orientation for the lower Paleozoic rocks using the bedding and cleavage attitudes measured during the day. Note that this can be done using a stereonet.

Question of the day:

- How to separate out the early Paleozoic (Wuyi-Yunkai) deformation from the "Indosinian" deformation in this region?
- The Wuyi-Yunkai Orogeny finished at ca. 420 Ma (see Section 1.6). Why Silurian strata are missing from this region?

Xinchang Dafosi (Big Budda Temple) billt on
massive Cretaceous red sandstones and tuff layers (Zheng-Xiang Li)

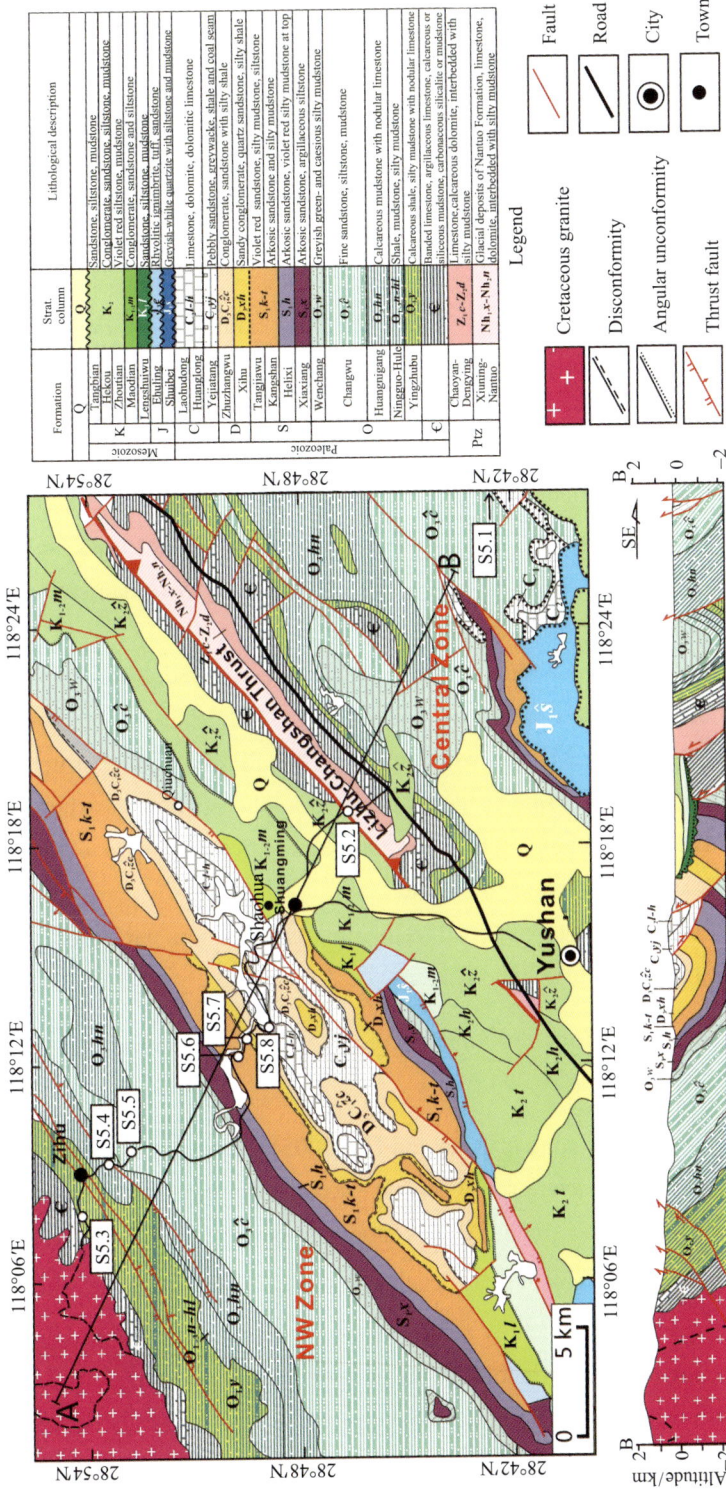

Figure 2.24   Geological map of the Yushan region and Day 5 field stops (after 1:250,000 geological map)

➤Day 5. Early Paleozoic strata along the Sanqing-Shaohua section, and deformation contrast between the two sides of the Lizhu-Changshan Thrust: Onset and lateral extent of the Wuyi-Yunkai Orogeny

Aim: To see how the sedimentary environment changed from a carbonate platform to a foreland basin setting (coarsening upward) during the Early Paleozoic due to the onset of the Wuyi-Yunkai Orogeny, and how the environment changed back to carbonate platform after the Wuyi-Yunkai Orogeny; To analyse the northern boundary of the Wuyi-Yunkai Orogen, and examine how an orogenic event is recorded differently at different parts of a continent.

Contents: Compare deformation styles across the Lizhu-Changshan Thrust (boundary fault between the Central and NW structural zones); Examine along the Sanqing-Shaohua section Cambro–Ordovician carbonate rocks, Ordovician–Carboniferous clastic rocks and sedimentary structures, disconformity between the Early Silurian and Late Devonian strata, and Carboniferous carbonates.

The Sanqing-Shaohua section is within the northwest structural zone of Xiao and He (2005) (Figure 2.18(b), (c)) and features a series of NW-verging, likely thin-skinned thrust sheets (Figure 2.18(b), (c) and Figure 2.24). We will see that in the SE and central structural zones (Stops 5.1 and 5.2, plus what we saw in day 4) to the southeast of the Lizhu-Changshan Thrust, Wuyi-Yunkai (Early Paleozoic) deformation appears to dominate. We will see whether deformation in the NW structural zone is the same.

Note: We will be working along a busy road (remember that safety is paramount) – extreme caution is needed when you are out of the vehicle!

Stop 5.1 Strong pre-Carboniferous deformation recorded by Ordovician strata east of the Gengdu village (28°42.466′N, 118°28.094′E)

We are in the central structural zone of Xiao and He (2005). On the geological map (Figure 2.18(a)), there is a clear angular unconformity between strongly deformed Ordovician strata and more gently folded Carboniferous strata. At this stop we examine: ① the Late Ordovician Wenchang Formation pelitic rocks with well-developed sub-vertically cleavage; and ② Early Carboniferous Yejiatang Formation pebbly sandstones with much shallower beddings and no cleavage. The contact between the two formations is not exposed, and further west along the road we can see disturbance of bedding in the Carboniferous strata, possibly by a small local fault. This angular unconformity, featuring contrasting deformation styles and sedimentary environments, and a long time gap between the two units, signifies the early Paleozoic

Wuyi-Yunkai Orogeny here.

Try to identify bedding and take measurements of both the bedding planes and the cleavage.

Stop 5.2   Small thrust duplex in Neoproterozoic strata just south of the Lizhu-Changshan Thrust (28°46.675′N, 118°19.089′E)

The Lizhu-Changshan Thrust is the boundary fault dividing the NW zone from the central zone in the structural map of Xiao and He (2005) (Figure 2.18(b) and Figure 2.24). The main thrust is not easily visible on outcrops, but here we see a small thrust system in the steeply SE-dipping Neoproterozoic clastic rocks in the hanging wall of the thrust fault. The thrust shows top-to-the-north sense of movement, similar to that of the main thrust. The low-angle thrust plane has a number of ramps, each leading to a smaller, bedding-parallel thrust surface. We will discuss the possible age(s) of the thrust movements.

After this quick stop, we cross the Lizhu-Changshan Thrust to the NW structural zone.

Stop 5.3   Cambro-Ordovician carbonates (28°54.123′N, 118°07.609′E)

Cambrian carbonate outcrops are visible further north from this stop, but not very good. At this stop we will see siliceous/carbonaceous banded limestone (Figure 2.25(a), (b)). The unit here was mapped as the Late Cambrian Huayanci Formation on the old 1:200,000 geological map, but on a newer map it is shown as the Early Ordovician Yinzhupu Formation (Figure 2.24; The lithology here appears to match the description for the Cambrian strata better – further paleontological investigations and regional stratigraphic correlations are required to resolve this).

Stop 5.4   Mid-Ordovician nodular limestone (28°53.407′N, 118°09.202′E)

Walk upstream along the riverbed, we will see mid-Ordovician (the Huangnigang Formation) nodular limestone outcrops (Figure 2.25(c)). In places, NE-trending and sub-vertical cleavage is well-developed (Figure 2.25(d)). Think about the likely age for the cleavage development. (There are plenty of nice looking river pebbles of dominantly Cambro-Ordovician rocks along the riverbed).

Stop 5.5   Late Ordovician Changwu Formation fine-grained sandstones and a small thrust fault (28°52.741′N, 118°09.587′E)

Thick to massive bedded, fine-grain sandstones and siltstones, with sedimentary structures such as ripple marks, cross-laminations (Figure 2.26(a)), and cross-beds. The massive thickness of the clastic rocks in this succession (before and after this stop), and their shallow-water nature, suggest that there was a steady growth of sedimentary accommodation as well as plentiful sedimentary sources, consistent with a foreland basin setting.

Figure 2.25   Early Ordovician banded limestone ((a), (b); Stop 5.3) and mid-Ordovician nodular muddy limestone((c), (d); Stop 5.4)

Figure 2.26   Sedimentary structures and a small thrust in the Late Ordovician Changwu Formation

At this outcrop, there is also a small SE-dipping thrust fault (Figure 2.26(b))—we can discuss the likely age of this fault in view of the local tectonostratigraphic record. For regional structures, see Figures 2.18(b), (c) and 2.24.

Stop 5.6   Pinkish Silurian sandstones–the last of the foreland basin deposits (28°49.920′N, 118°12.430′E)

We will examine here the uppermost unit of the O—S foreland basin deposits, mapped as Early Silurian age (Figure 2.24). The grain size of the foreland basin deposits appears to have become coarser, and the color is now becoming pinkish, possibly reflecting increasing fluvial influence as the orogenic front moved closer.

Stop 5.7   Silurian–Devonian disconformity, signifying the impact of the Wuyi-Yunkai Orogeny here (28°49.657′N, 118°12.943′E)

We will walk up the section to see the change in the post-orogenic sedimentation. Sediments below the disconformity are reddish in color, possibly reflecting fluvial environments toward the later stages of the foreland basin environment (Figure 1.20). Much of the continent remained exposed to surface erosion from the end of the Wuyi-Yunkai Orogeny until the Devonian-Carboniferous marine transgression, starting with basal conglomerates and quartz sandstones (Devonian sediments as in this stop; Figure 1.22(a), (b)) and finishing with platform carbonates (next stop; Figure 1.22(c), (d)).

Things to do: See whether you can find the S/D disconformity by checking layer by layer along the roadside outcrop–digging is a must. Take note of changes in lithology and texture as you go, and also observe and measure the bedding attitude along the way. You should finish with a simple cross-section on your notebook.

Here we see Early Silurian foreland basin deposits disconformably overlain (i.e., in parallel unconformity) by Late Devonian conglomeratic coarse sandstones. There is no angular unconformity here as what we saw at stops 4.4 and 5.1, and the age span of the sedimentary hiatus here is much smaller too.

Q1: Did the Wuyi-Yunkai Orogeny cause any deformation in this region?

Q2: What are the changes in composition and texture across the disconformity? Why?

Q3: Why Silurian and Devonian rocks are found in this region, but not in the Jiangshan region?

Stop 5.8   Late Carboniferous limestone outside the entrance to the Tianliang park (28°49.017′N, 118°13.132′E)

During the Devonian–Carboniferous time (between Stop 5.7 and this stop) South China recorded a marine transgression. By the Late Carboniferous, almost the entire continent was covered by platform carbonates (Figure 1.22 and Figure 2.27).

Figure 2.27   Late Carboniferous Huanglong Formation massive limestone

Exercises of the day:

• Draw a tectonostratigraphic diagram for the region, and compare it with that of the Jianshan region just tens of kilometres to the southeast.

• Now, use the four tectostratigraphic diagrams to compile a regional time-space diagram, in the order of, from left to right, Yushan, Jiangshan, Shuangxiwu, and Chencai. Identify the differences and similarities, and interpret them in terms of the tectonic history of eastern South China.

• Work in groups to prepare a PowerPoint presentation of the tectonic history of eastern South China using what we have seen and learnt in the past days, and the time-space diagrams that you developed during the trip.

Question of the day:

• What may have caused the dramatic difference between the tectonostratigraphy of the Jianshan and Yushan regions?

➤Day 6. Structural and stratigraphic records of the Permian—Triassic Indosinian Orogeny, and post-orogenic basin and magmatic developments in NE Jiangxi

Aim: See how geological history almost repeated in the region during the Permian–Cretaceous time (in comparison with the Ordovician–Devonian orogenic, basin and magmatic records that we observed in the previous days).

Contents*: The Yingjia section–Permo-Carboniferous platform carbonates gradually changed to late Permian–Triassic clastic rocks with coal seams (foreland basin deposits); angular unconformity between Jurassic strata (post-orogenic) and older rocks; Late Jurassic — Cretaceous volcanic deposits and redbeds.

Stop 6.1　Strongly deformed Early to mid-Permian (Qixia-Maokou formations) limestone (28°14.282′N, 118°00.110′E)

Walking along the bottom of a gully, we will see outcrops of medium layered limestone of mid-Permian age–no sign of an approaching orogen here yet (Figure 2.29). Good outcrops are also available on top of the hills. Late Carboniferous (to Early Permian?) Chuanshan Formation massive limestone outcrops are also found near the freeway entry at Shangrao (28°30.213′N, 117°58.147′E) which we may drive pass during the day.

Stop 6.2　Mid-Permian (upper Chetou Formation) fine-grained sandstones (28°14.120′N, 118°1.488′E)

Poorly outcropped mid-Permian fine-grained sandstone along the dirt driveway to a farmhouse on the roadside. The occurrence of clastic rocks signifies the region being converted from a carbonate platform environment to a foreland basin environment, indicating the approaching of the Indosinian Orogen from the southeast (see Section 1.7).

Stop 6.3　Strongly deformed Late Permian (the Leping Formation) sandstones, and carbonaceous siltstones and shale, just below the angular unconformity of the Indosinian Orogeny here (28°14.285′N, 118°2.448′E)

Have a look at Figure 2.28 now, you will notice that Jurassic strata in this region lie atop (like "floating" on) more strongly deformed pre-Jurassic rocks – they have been interpreted to be products of different tectonic regimes.

---

　* Finding a good traverse for the purpose of examining the Indosinian (Mid-Permian to Late Triassic) orogenic cycle is not easy along the corridor of this field excursion. The selected stops (Figure 2.28(c)) cover the main elements required for the purpose, but the outcrop conditions are not always great. Some alternative stops are shown on Figure 2.28(b), with selected photos shown for comparable stops.

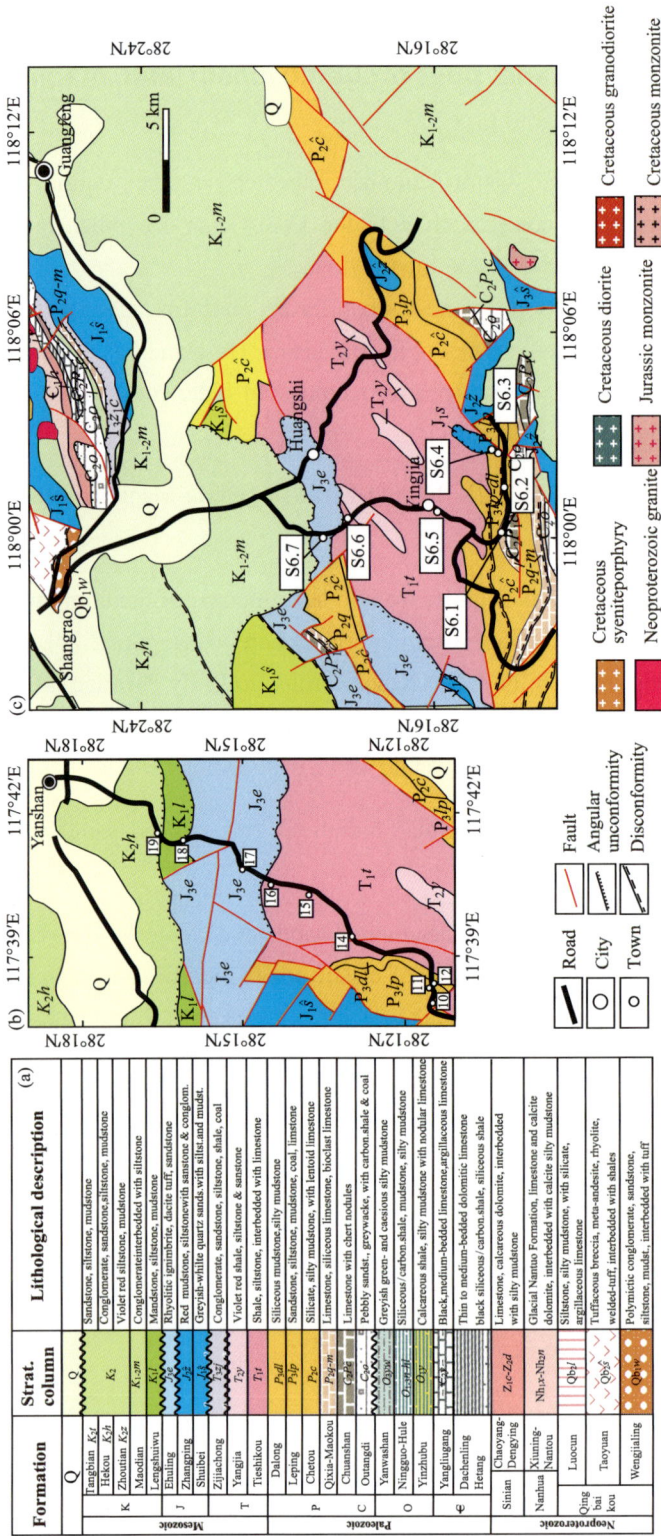

Figure 2.28   Simplified geological maps of the Yingjia and southern Yanshan regions, and field stops for Day 6 (after 1:250,000 geological map). For locations of the two maps ((b), (c)) see field locations for Day 6 in Figure 2.1 (marked as 6 and 6′).

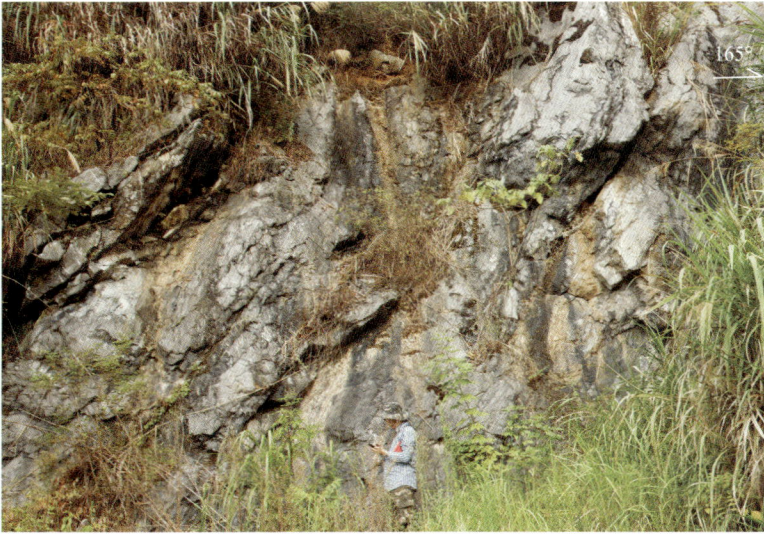

Figure 2.29   Early to mid-Permian Chuanshan-Qixia-Maokou formations limestone

At this stop, we see Late Permian sandstones, carbonaceous siltstones and black shale exposed just below an angular unconformity with the overlying Early Jurassic clastic rocks (Stop 6.4). This angular unconformity signifies the Indosinian Orogeny here. The pre-Jurassic rocks are much more strongly deformed – take measurements of bedding planes here, and compare them with the J_1 bedding attitude above the unconformity at the next stop.

Coal seams were widely developed both along the SE margin of the Indosinian foreland basin that migrated from along the southeast coastal region in the Permian (Figure 1.24(b)), to the Sichuan basin by the end of the Triassic (Figure 1.24(d)), and within the Late Triassic-Early Jurassic sag basin (Figure 1.24(d), (e)). Coal seams are present in the Leping Formation in this region, and are mined by local farmers, but we see only carbonaceous black shale at this stop.

Figure 2.30 shows photographs taken at equivalent (but better exposed) supplementary stops #10 and #12 as marked in Figure 2.28(b).

Stop 6.4   Structure and composition of Early Jurassic (the Shuibei Formation) clastic rocks just above the Indosinian angular unconformity (28°14.418′N, 118°2.450′E)

Along a dirt track west of the river, we see poorly exposed arkosic sandstones, quartz sandstones (Figure 2.31(a)) and black shale that lie just above the angular unconformity. Take a bedding measurement here and compare it with that from the nearby Stop 6.3 in the underlying rocks. Better sandstone outcrops with cross-beds can be seen east of the river.

The Late Triassic–Early Jurassic strata in central and southeastern South China (Figure 1.24(d),(e)) sit unconformably on top of Indosinian-deformed strata or igneous/metamorphic rocks. Sediments in this basin have been mapped to be thousands of meters thick that experienced only mild deformation (commonly tilting only, rarely seen being folded). This basin did not attract much attention until Li and Li (2007) interpreted it as a sag basin formed by the drag force of an eclogised oceanic plateau that was flat-subducted under southeastern South China during the Indosinian Orogeny (see Section 1.7). Can you think of an alternative explanation for both the broad (ca. 1300 km) Indosinian Orogen (or Huanan Orogen?), and this almost equally broad sag basin?

Figure 2.30 (a) Late Permian Leping Formation strongly deformed sandstones and black shale at supplementary Stop 12 (28°11.577′N, 117°38.482′E). (b)–(d) Calcareous mudstone, thin muddy limestone and siltstone with axial cleavage (b) and cross laminations (d) at supplementary Stop 10 (28°11.500′N, 117°38.092′E; see Figure 2.28(b) for positions of the supplementary stops)

Figure 2.31   Early Jurassic Shuibei Formation quartz sandstones (a) and pebbly sandstones that contain carbonaceous clasts of unknown age (conglomerates or intraclasts?)

Stop 6.5   Structure and lithology of Early Triassic (the Tiekou Formation) siltstone and silty mudstone (28°15.893′N, 118°00.773′E)

Back to the strongly folded Permo-Triassic succession below the Indosinian angular unconformity, we see here sub-vertical, strongly weathered Early Triassic Tiekou Formation, featuring thin-bedded, calcareous(?) siltstone and mudstone (Figure 2.32(a)). At supplementary (Stop 15 in Figure 2.28(b)) there are well developed cross laminations (Figure 2.32(b)). Further south on the roadside there is a small outcrop showing a chevron fold in the calcareous siltstone.

Figure 2.32   Sub-vertical Early Triassic Tiekou Formation calcareous siltstone and mudstone (a). At supplementary Stop 15 (Figure 2.28(b)) there are well developed cross laminations (b)

Stop 6.6   Kink folding in the top section of the Early Triassic (the Tieshikou Formation) just below the angular unconformity (28°18.368′N, 118°00.407′E)

If the outcrop is not totally blocked off by new farmhouses, we should see here kink folds in Early Triassic, light grey, slaty siltstone(?). Draw a sketch of the outcrop and take some S_0/S_1 and kink fold measurements. Compare both the rock type (thus deposition environment) and deformation intensity and styles with those at the next stop, which was formed after the Indosinian Orogeny.

Stop 6.7   Late Jurassic massive welded tuff with cooling joints that lies above the Indosinian angular unconformity (28°19.002′N, 117°59.970′E)

Both the composition and the structure of this Late Jurassic unit (the Ehuling Formation) are in sharp contrast to the underlying strata that we observed earlier in the day below the Indosinian angular unconformity. The formation is dominantly volcanic in composition with shallowly-dipping bedding toward NNE. The volcanic and volcaniclastic rocks in this region are parts of the post-orogenic large magmatic province in central and southeastern South China (Figure 1.25).

Figure 2.33   Kink folds in the Early Triassic strata that developed a strong bedding-parallel cleavage

Figure 2.34   Late Jurassic massive Ehuling Formation welded tuff with cooling joints

Note that the volcanic rocks here are dominantly of Late Jurassic age, whereas those further to the southeast (i.e. Day 3 stops) are dominantly of Cretaceous age. Zhou and Li (2000) interpreted this younging-to-the-southeast trend as a result of changes in Palaeo-Pacific subduction angle from shallow to steep, whereas Li and Li (2007) interpreted it as a result of slab foundering that started from the southern Hunan-northern Guangdong-southwestern Jiangxi region (Section 1.7, Figure 1.25).

Exercises of the day:
  • Group presentations of the tectonic evolution of the region.
Question of the day:
  • Was the Indosinian Orogeny related to the subduction of the Paleo-Pacific oceanic plate, or the collision of the Indochina Block with South China? Why?

Waterfall in the Yandangshan Cretaceous volcanic geopark, east
Zhejiang. Note the presence of regional tilting but not folding. (Zheng-Xiang Li)

➤Day 7. Tianli Schists, Neoproterozoic rift magmatism and the Zhang-shudun serpentinite complex, NE Jiangxi

Aim: To examine more evidence for an active southern margin of the Yangtze Block during late Mesoproterozoic to earliest Neoproterozoic (the Sibao orogenesis), and mid-Neoproterozoic continental rifting.

Contents: The Wongjialing section: the Tianli Schists and the overlying volcaniclastic rocks; the Shilong section: 825 Ma bimodal magmatism; the ca. 1000 Ma Zhangshudun serpentinite.

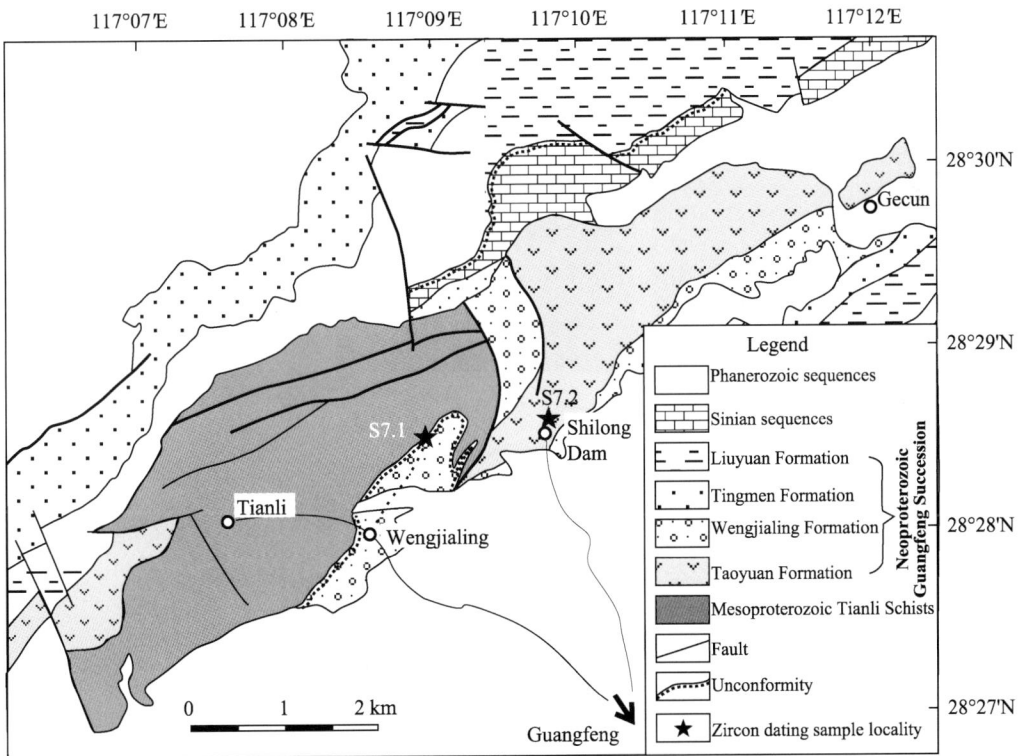

Figure 2.35   A simplified geological map of the Guangfeng region (after Li W X et al., 2008b)

Stop 7.1   Tianli Schists and angular unconformity (28°28.301′N, 118°08.832′E)

The Tianli Schists underlying the Neoproterozoic rift successions near Guangfeng (Figure 2.35 and Figure 2.36) are the only known metamorphic rocks in eastern South China that were metamorphosed during the Sibao Orogeny. It was possibly on the same continental margin (the southern Yangtze margin) as the Shuangxiwu arc system (Figure 1.1, Figure1.5, Figure

2.35; Li et al., 2007). The "angular" unconformity here represents a deposition hiatus of hundreds of millions of years, as well as a great contrast in metamorphic grades: non-metamorphosed mid-Neoproterozoic rift succession (with basal conglomerates) directly overlying ca. 1 Ga higher-greenschist metamorphic rocks (Figure 2.36).

U-Pb zircon provenance ages indicate that the protolith of the Tianli Schists was a clastic sedimentary succession most likely derived from the Yangtze Block (Figure 2.37; Li et al., 2007). The depositional age of the protolith is younger than 1530 Ma, as constrained by the youngest detrital zircon grains, but is older than 1040 Ma as constrained by the oldest ^{40}Ar/^{39}Ar muscovite ages (Figure 2.38). As shown in Figure 2.38, the Tianli Schists appear to have been metamorphosed at 1042–1015 Ma, and recorded tectonic reactivations at 968 ± 4 Ma and 942 ± 8 Ma, all considered parts of the Sibao Orogeny (Figure 1.5; Li et al., 2007).

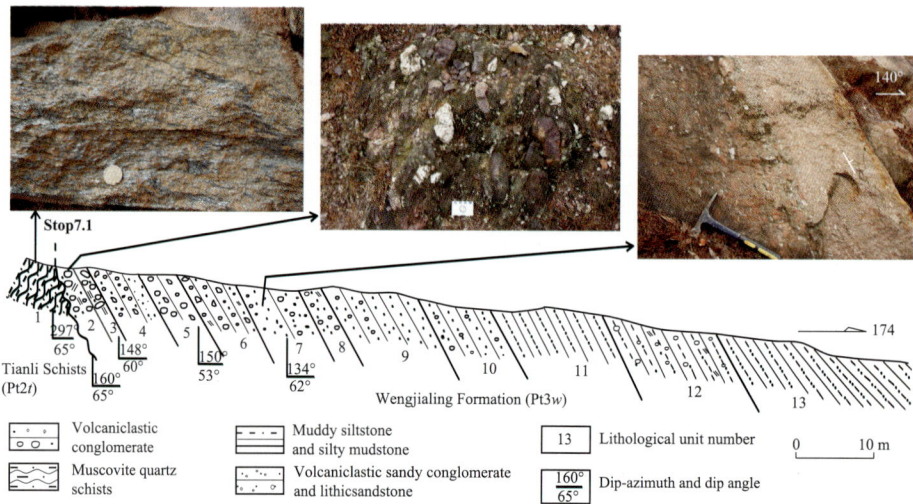

Figure 2.36   A cross-section at Stop 6.1 (after Wang et al., 2003)

At this stop, we see the non-metamorphosed Neoproterozoic Wengjialing Formation basal conglomerates deposited directly on the Tianli Schists (Figure 2.36). The Wengjialing Formation consists of purplish volcaniclastic conglomerates and volcaniclastic sandstones. The gravels are dominantly of rhyolite and quartz, with occasional basaltic gravels. Fragments of muscovite schists, likely from the underlying Tianli Schists (Figure 2.36), are abundant.   Up-section in the Wengjialing Formation, the lithology grades into lithic sandstones, siltstones, muddy siltstones and mudstones. Vertically, this formation consists of an overall fining-upward sequence containing several fining-upward cycles (Figure

2.36; Wang et al., 2003).

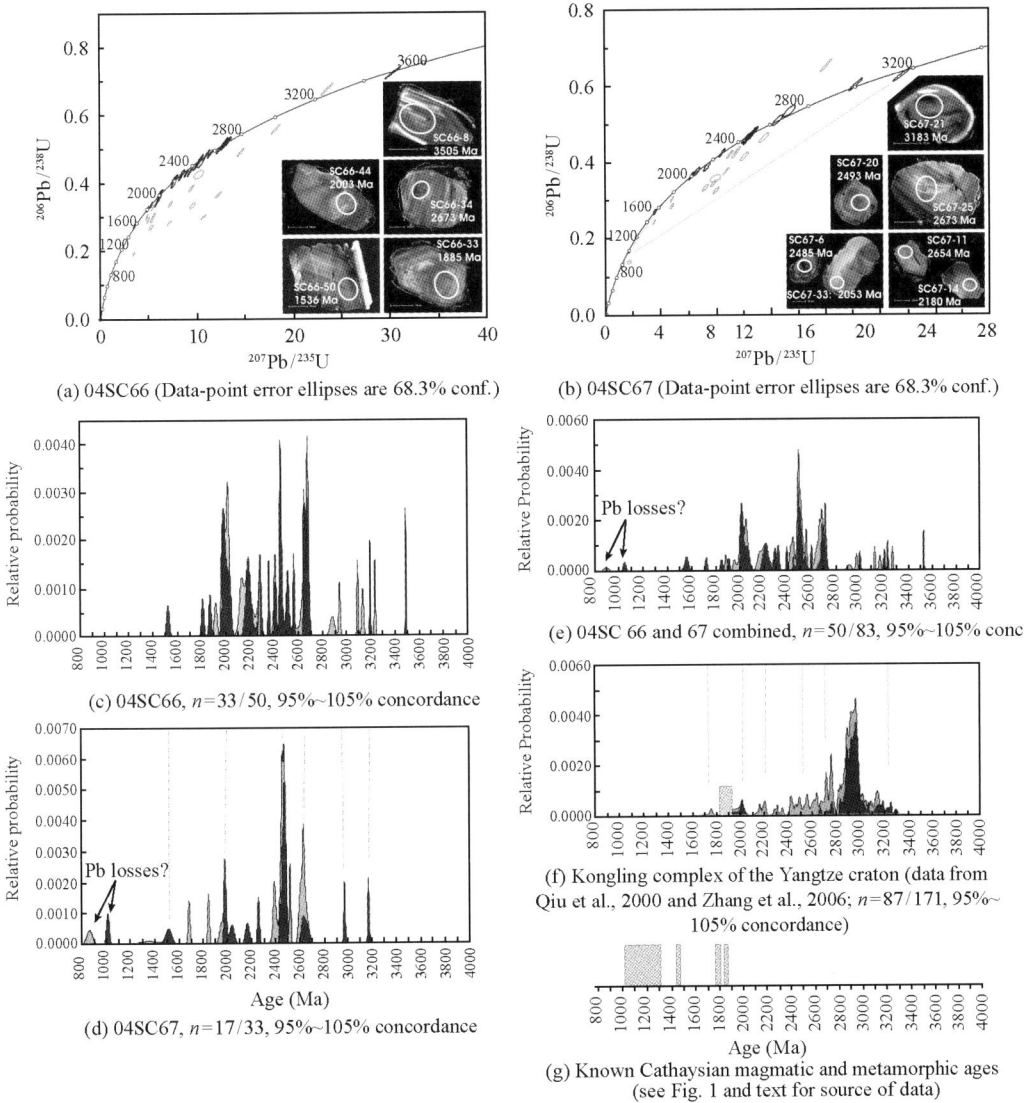

(a) 04SC66 (Data-point error ellipses are 68.3% conf.)

(b) 04SC67 (Data-point error ellipses are 68.3% conf.)

(c) 04SC66, $n=33/50$, 95%~105% concordance

(d) 04SC67, $n=17/33$, 95%~105% concordance

(e) 04SC 66 and 67 combined, $n=50/83$, 95%~105% conc

(f) Kongling complex of the Yangtze craton (data from Qiu et al., 2000 and Zhang et al., 2006; $n=87/171$, 95%~105% concordance)

(g) Known Cathaysian magmatic and metamorphic ages (see Fig. 1 and text for source of data)

Figure 2.37   Sedimentary detrital age data (SHRIMP) indicate a Yangtze provenance for the protolith of the Tianli Schists (Li et al., 2007)

Stop 7.2   Ca. 825 Ma Taoyuan Formation bimodal volcanic rocks (28°28.621′N, 118°09.741′E) and the Wengjialing Formation volcaniclastic rocks (28°28.812′N, 118°09.604′E – the contact between the Taoyuan and Wengjialing formations) along the small road on the bank of the Shilong Reservoir

Neoproterozoic rift successions in the Guangfeng region consist of four sedimentary facies

85

associations: ① the Taoyuan Formation continental bimodal volcanic rocks, ② the Wengjialing Formaiton alluvial and fluvial volcaniclastic rocks and lacustrine fine-grained clastic rocks, ③ the Tingmen Formation fluvial arkosic sandstones, and the Liuyuan Formation littoral to shallow-marine quartz sandstones, and ④ the Nantuo tillites. The rift succession is overlain by platform carbonate rocks (the Chaoyang and Dengying formations; Wang et al., 2003). At this section, we examine the Taoyuan Formation bimodal volcanic rocks and the Wengjialing Formation volcaniclastic rocks.

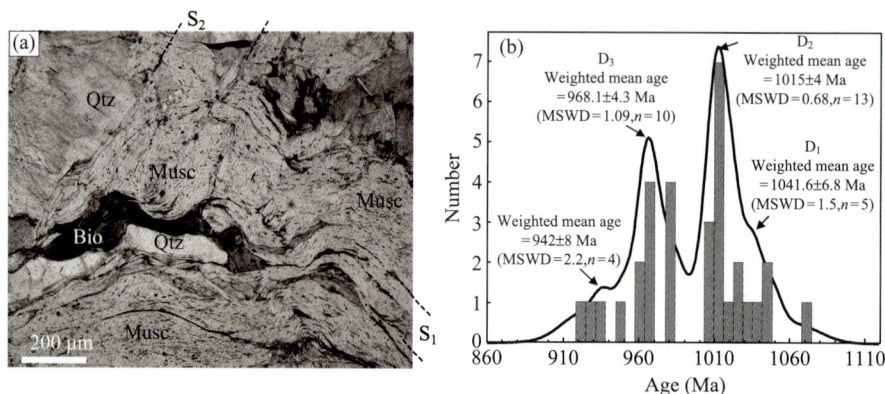

Figure 2.38 (a) Optical photomicrograph of deformed white micas in Tianli quartz mica schist sample 04SC67 showing S_1 and S_2 foliations. Musc = muscovite, Bio = biotite, and Qtz = quartz. (b) Histogram (grey boxes) and cumulative probability (solid black line) plots of the combined $^{40}Ar/^{39}Ar$ UV laser spot mica ages from Tianli Schists samples (Li et al., 2007)

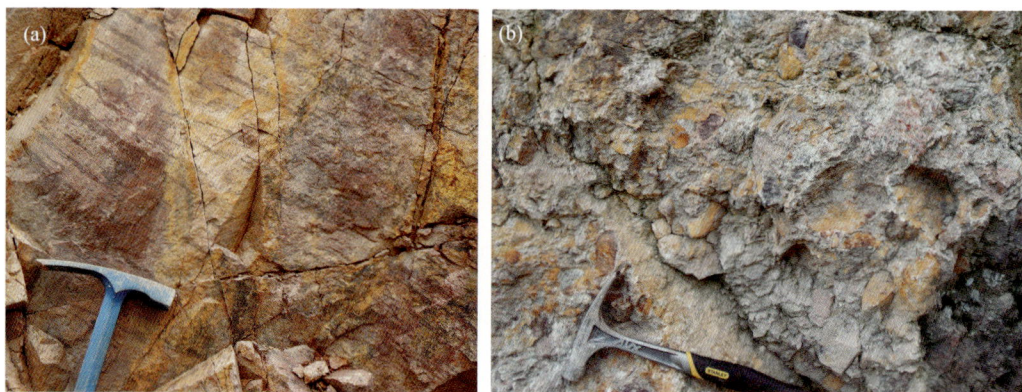

Figure 2.39 Neoproterozoic Taoyuan Formation ignimbrite with rhyolitic texture (a) and Wengjialing Formation volcanic conglomerate (b) at Shilong Dam

The Taoyuan Formation has a total thickness of ca.130 m and comprises basalts, dacites/rhyolites intercalated with volcaniclastic rocks, ignimbrite (Figure 2.39(a)) and tuff. Li W X et al. (2008b) dated a rhyolite member of the Taoyuan Formation at 827 ± 14 Ma. Their geochemical analyses revealed that the volcanic rocks consist of alkaline basalts, andesites and peraluminous rhyolites.

Despite showing Nb-Ta depletion relative to La and Th, the alkaline basalts are characterized by highly positive Nd(T) values (+3.1 to +6.0), relatively high TiO_2 and Nb contents, high Zr/Y and super-chondritic Nb/Ta ratios, suggesting their derivation from a slab melt-metasomatised subcontinental lithospheric mantle source in an intracontinental rifting setting (Li W X et al., 2008b). These authors concluded that the Guangfeng volcanic suite is a magmatic response of variant levels of continental lithosphere (including lithospheric mantle and the lower-middle to upper crust) to the mid-Neoproterozoic intracontinental rifting.

For more general discussions regarding Neoproterozoic rift magmatism and possible links to mantle plume, see Section 1.5 and references therein.

Disconformably overlying the Taoyuan Formation is the Wengjialing Formation with a thickness of ca. 400 m, consisting of volcaniclastic conglomerate (Figure2.39(b)), pebbly sandstone, muddy-siltstone and mudstone.

Stop 7.3   The Zhangshudun serpentinite complex (28°32.734′N, 117°26.736′E)

The NE Jiangxi ophiolite belt (Zhou, 1989; see position shown in Figure 1.1) has been regarded as being formed in either a back-arc basin or an intra-arc basin along the southern margin of the Yangtze Block before it was joined with the Cathaysia Block in earliest Neoproterozoic (Li et al., 1997; Li et al., 2007; Li W X et al., 2008a) (Figure 1.5). The serpentinite seen at this stop, widely interpreted as part of an ophiolite complex, was first dated at 1034 ± 24 Ma by Chen et al. (1991; Sm-Nd mineral internal isochron age). Li W X et al. (work in-progress) recently obtained a SIMS U-Pb zircon age of 988 ± 13 Ma for cumulus gabbro in the ophiolite complex, which is consistent with the Sm-Nd age.

At the site (S7.3 in Figure 2.40), there is a layer with pillow-like structures (Figure 2.41). The origin of such structures is currently under investigation. Li W X et al. (work in-progress) preliminarily interpreted the rocks as "rodingite", likely formed through interactions between Ca- and Si-enriched fluids and mafic dikes or other rocks in the mafic-ultrabasic complex when the peridotites were serpentinized.

The "country rocks" for the serpentinite complexes found in the region has been mapped as the "Zhangcun Group" in recent geological maps (Figure 2.40), but the nature and ages of these rocks, as well as their deformation history, have not been well studied yet.

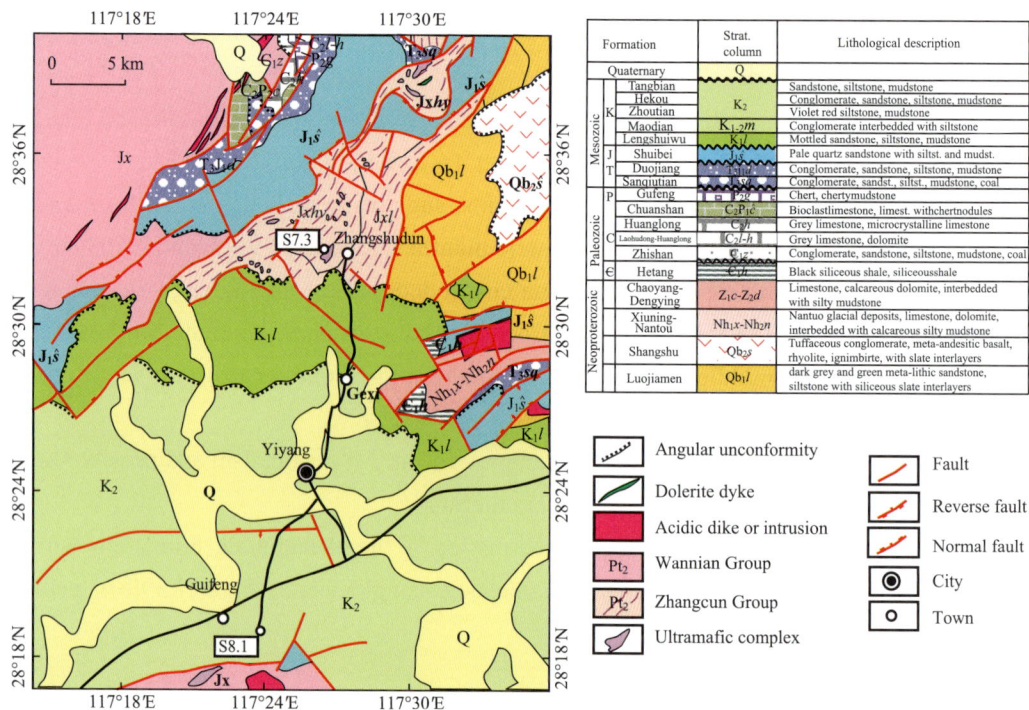

Formation		Strat. column	Lithological description
Quaternary		Q	
Mesozoic K	Tangbian		Sandstone, siltstone, mudstone
	Hekou	K_2	Conglomerate, sandstone, siltstone, mudstone
	Zhoutian		Violet red siltstone, mudstone
	Maodian	$K_{1-2}m$	Conglomerate interbedded with siltstone
J	Lengshuiwu	$K_1 l$	Mottled sandstone, siltstone, mudstone
T	Shuibei		Pale quartz sandstone with siltst. and mudst.
	Duojiang	$J_1 s\hat{}$	Conglomerate, sandstone, siltstone, mudstone
	Sanqiutian	$J_1 x$	Conglomerate, sandst., siltst., mudstone, coal
Paleozoic P	Gufeng	$P_2 g$	Chert, chertymudstone
	Chuanshan	$C_2 P_1 \hat{c}$	Bioclastlimestone, limest. withchertnodules
C	Huanglong	$C_2 h$	Grey limestone, microcrystalline limestone
	Laohudong-Huanglong	$C_2 l-h$	Grey limestone, dolomite
	Zhishan	$\in_{3}\hat{z}$	Conglomerate, sandstone, siltstone, mudstone, coal
€	Hetang	$\in_1 h$	Black siliceous shale, siliceousshale
Neoproterozoic	Chaoyang-Dengying	$Z_1 c-Z_2 d$	Limestone, calcareous dolomite, interbedded with silty mudstone
	Xiuning-Nantou	$Nh_1 x-Nh_2 n$	Nantuo glacial deposits, limestone, dolomite, interbedded with calcareous silty mudstone
	Shangshu	$Qb_2 s$	Tuffaceous conglomerate, meta-andesitic basalt, rhyolite, ignimbirte, with slate interlayers
	Luojiamen	$Qb_1 l$	dark grey and green meta-lithic sandstone, siltstone with siliceous slate interlayers

Legend:
- Angular unconformity
- Dolerite dyke
- Acidic dike or intrusion
- Pt_2 Wannian Group
- Pt_2 Zhangcun Group
- Ultramafic complex
- Fault
- Reverse fault
- Normal fault
- City
- Town

Figure 2.40 Geological map of the Zhangshudun-Yiyang region (after 1:250,000 geological maps)

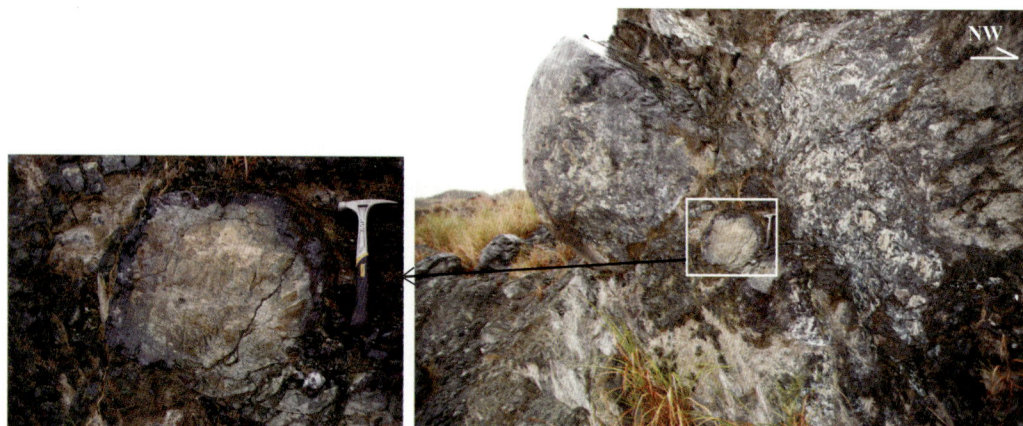

Figure 2.41 Pillow-like structures in the Zhangshudun serpentinite complex

Question of the day:

- Which sections of the ophiolite sequence the observed rocks at Zhangshudun (S7.3) may represent?

➤Day 8. Guifeng Geopark: Cretaceous redbeds and Danxia landform

Extensive red continental deposits were developed over a wide area of South China during the Cretaceous. However, what controlled the formation of such redbeds is still unclear. Preliminary work indicates that widespread extensional or transtensional basins may have controlled their distribution and thickness (i.e, similar to the basin-and-range province in western North America), but much more work is required to provide a clearer picture of what happened during that time interval.

In the morning of Day 8 (S8.1 in Figure 2.40; Figure 4.22), we will examine the composition and sedimentary structures of the Cretaceous sediments, and the unique landscape they form. Note the significantly reduced volcanic input to the Cretaceous deposits here, in comparison with what we saw in the Xinchang region (Day 3).

Enjoy the afternoon on an easy ride home.

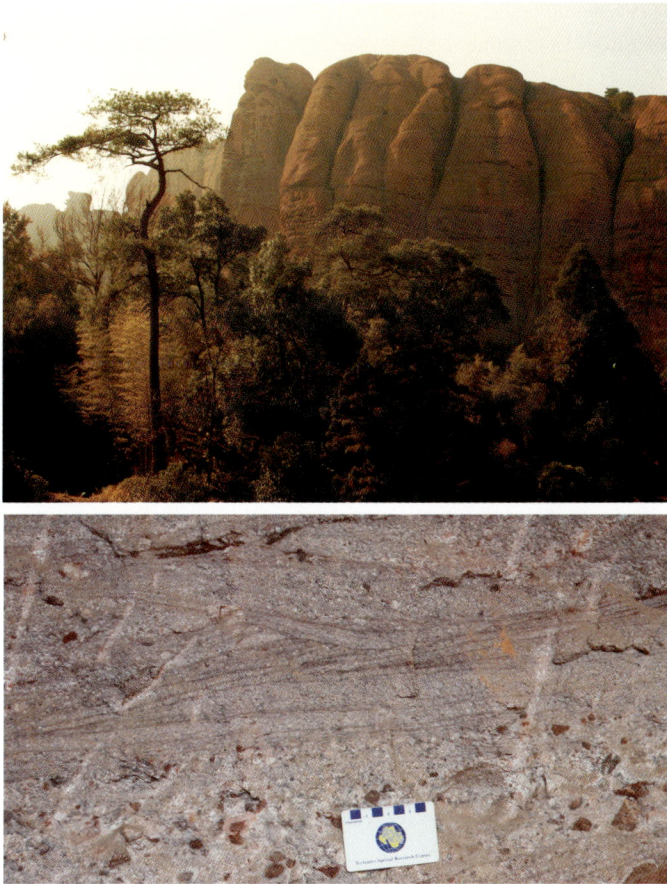

Figure 2.42   Massive Late Cretaceous continental conglomeratic sandstones in the Guifeng (turtle peak) Danxia geopark. The strata experienced block tilting, but not folding

References

BOGAMR (Bureau of Geology and Mineral Resources, Zhejiang Province). 1989. Regional Geology of zhejiang Province Gedegical Memoirs, 11. Beijing: Geological Publishing House.

Chen J, Foland K A, Xing F, et al. 1991. Magmatism along the southeast margin of the Yangtze Block: precambrian collision of the Yangtze and Cathysia block of China. Geology, 19: 815–818.

Chen Z, Xing G, Guo K, et al. 2009. Petrogenesis of keratophyes in the Pingshui Group, Zhejiang: constraints from zircon U-Pb ages and Hf isotopes. Chinese Science Bulletin, 54: 1570–1578.

Dong C W, Yang S F, Tang L M, et al. 2008. Petrology and geochemistry of the Xinchang composite Igneous complexes, Zhejiang and their geological implication. Geological Journal of China Universities, 14: 365–376.

Eby, G N. 1992. Chemical subdivision of the A-type granitoids: petrogenetic and tectonic implications. Geology, 20: 641–644.

Gilder S A, Keller G R, Luo M, et al. 1991. Timing and spatial distribution of rifting in China. Tectonophysics, 197: 225–243.

Jahn B M, Zhou X H, Li J L. 1990. Formation and tectonic evolution of Southeastern China and Taiwan: isotopic and geochemical constraints. Tectonophysics, 183: 145–160.

Kong X, Li Z, Feng B, et al. 1995. The Precambrian Geology of Chencai Region in Zhejiang Province. Precambrian Geology, 7. Beijing:Geological Publishing House(in Chinese with English summary).

Li J. 1993. Tectonic framework and evolution of southeastern China. Journal of Southeast Asian Earth Science, 8: 219–223.

Li W X, Li X H, Li Z X, 2005. Neoproterozoic bimodal magmatism in the Cathaysia Block of South China and its tectonic significance. Precambrian Research, 136: 51–66.

Li W X, Li X H, Li Z X, et al. 2008a. Obduction-type granites within the NE Jiangxi Ophiolite: implications for the final amalgamation between the Yangtze and Cathaysia Blocks. Gondwana Research, 13: 288–301.

Li W X, Li X H, Li Z X. 2008b. Middle Neoproterozoic syn-rifting volcanic rocks in Guangfeng, South China: petrogenesis and tectonic significance. Geological Magazine, 145: 475–489.

Li X H, Zhao J X, McCulloch M T, et al. 1997. Geochemical and Sm-Nd isotopic study of Neoproterozoic ophiolites from southeastern China: petrogenesis and tectonic implications. Precambrian Research, 81: 129–144.

Li X H, Li W X, Li Z X, et al. 2008. 850-790 Ma bimodal volcanic and intrusive rocks in northern Zhejiang, South China: a major episode of continental rift magmatism during the breakup of Rodinia. Lithos, 102: 341–357.

Li X H, Li W X, Li Z X, et al. 2009. Amalgamation between the Yangtze and Cathaysia Blocks in South China: Constraints from SHRIMP U-Pb zircon ages, geochemistry and Nd-Hf isotopes of the Shuangxiwu volcanic rocks. Precambrian Research, 174: 117–128.

Li Z X, Li X H, Kinny P D, et al. 2003a. Geochronology of Neoproterozoic syn-rift magmatism in the Yangtze Craton, South China and correlations with other continents: evidence for a mantle superplume that broke up Rodinia. Precambrian Research, 122: 85–109.

Li Z X, Li X H, Wang J, et al. 2003b. From Sibao Orogenesis to Nanhua Rifting: Late Precambrian Tectonic History of Eastern South China. Beijing:Geological Publishing House.

Li Z X, Wartho J A, Occhipinti S, et al. 2007. Early history of the eastern Sibao Orogen (South China) during the assembly of Rodinia: New mica $^{40}Ar/^{39}Ar$ dating and SHRIMP U-Pb detrital zircon provenance constraints. Precambrian Research, 159: 79–94.

Li Z X, Bogdanova S V, Collins A S, et al. 2008. Assembly, configuration, and break-up history of Rodinia: a synthesis. Precambrian Research, 160: 179–210.

Li Z X, Li X H, Wartho J A., Clark, C., Li, W.X., Zhang, C.L., Bao, C.M., 2010. Magmatic and metamorphic events during the early Paleozoic Wuyi-Yunkai orogeny, southeastern South China: new age constraints and pressure-temperature conditions. Geological Society of America bulletin, 122: 772–793.

Li Z X, Li X H, Chung S L, et al. 2012. Magmatic switch-on and switch-off along the South China continental margin since the Permian: transition from an Andean-type to a Western Pacific-type plate boundary. Tectonophysics, 532-535: 271–290.

Ma L F, Qiao X F, Min L R, Fan B X, Ding X Z. 2002. Geological atlas of China. Beijing: Geological Publishing House.

Pang C J, Krapež B, Li Z X, et al. 2014. Stratigraphic evolution of a Late Triassic to Early Jurassic intracontinental basin in Southeastern South China: A corsequence of flat-slab subduction? Sedimentary Geology, 302: 44–63.

Pearce J A. 1982. Trace element characteristics of lavas from destructive plate boundaries. *In*: Thorpe R S. New York: Andesites Wiley: 525–548.

Pearce J A. 1996. A User's Guide to Basalt discrimination diagrams. *In*: Wyman, D A., Trace Element Geochemistry of Volcanic Rocks: Applications for Massive Sulphide Exploration Geological Association of Canada. Short Course Notes, 12: 79–113.

Pearce J A, Harris N B W, Tindle A G 1984. Trace element discrimination diagrams for the tectonic interpretation of granitic rocks. Journal of Petrology, 25: 956–983.

Rollinson H R. 1993. Using Geochemical Data: Evaluation, Presentation, Interpretation. London: Longman Geochemistry Society.

Shervais J W. 1982. Ti-V plots and the petrogenesis of modern and ophiolitic lavas. Earth and Planetary Science Letters, 59: 101–118.

Shui T. 1987. Tectonic framework of the southeastern China continental basement. Scientia Sinica, (B30): 414–422

Vermeesch P. 2006. Tectonic discrimination diagrams revisited. Geochemistry, Geophysics, Geosystems 7: Q06017. doi:10.1029/2005GC001092.

Wang J, Li Z X. 2003. History of Neoproterozoic rift basins in South China: implications for Rodinia break-up. Precambrian Research, 122: 141–158.

Wang J, Li Z X, Li X H. 2003. Day 3 to Day 4 morning: The Neoproterozoic rift successions and their angular unconformable contacts with the Mesoproterozoic Tianli schists (metamorphosed by the Sibao orogeny) in Guangfeng, Jiangxi. *In*: Li Z X, Wang J, Li X H, et al. From Sibao Orogenesis to Nanhua Rifting: Late Precambrian Tectonic History of Eastern South China - An Overview and Field Guide. Beijing: Geological Publishing House: 63–74.

Wang Q, Wyman D A, Li Z X, et al. 2010. Petrology, geochronology and geochemistry of ca. 780 Ma A-type granites in South China: petrogenesis and implications for crustal growth during the breakup of the supercontinent Rodinia. Precambrian Research, 178: 185–208.

Whalen J B. Currie K L, Chappell BW. 1987. A-type granites: geochemical characteristics, discrimination and petrogenesis. Contributions to Mineralogy and Petrology, 95: 405–419.

Xiao W, He H. 2005. Early Mesozoic thrust tectonics of the northwest Zhejiang region (Southeast China). GSA Bulletin, 117: 945–961.

Ye M F, Li X H, Li W X, et al. 2007. SHRIMP zircon U-Pb geochronological and whole-rock geochemical evidence for an early Neoproterozoic Sibaoan magmatic arc along the southeastern margin of the Yangtze Block. Gondwana Research, 12: 144–156.

Zhou G. 1989. The discovery and significance of the northeastern Jiangxi Province ophiolite (NEJXO), its metamorphic peridotite and associated high temperature-high pressure metamorphic rocks. Journal of Southeast Asian Earth Science, 3: 237–247.

Zhou X, Zhu Y. 1993. Late Proterozoic collisional orogen and geosuture in southeastern China: Petrological evidence. Chinese Journal of Geochemistry, 12: 239–251.

Zhou X M, Li W X. 2000. Origin of Late Mesozoic igneous rocks in Southeastern China: implications for lithosphere subduction and underplating of mafic magmas. Tectonophysics, 326: 269–287.

Quote of the day:
You will not only be choosing a career, but also a lifestyle (Zheng-Xiang Li)

Landscape at Zhangjiajie, NW Hunan, developed on gently-dipping Devonian sandstones that experienced Indosinian thin-skinned thrusting (Zheng-Xiang Li)

APPENDIX

华南东部地质演化——八日野外考察指南[*]

李正祥[1,2]　　陈汉林[2]　　李献华[3]　　章凤奇[2]

1 澳大利亚研究理事会之核壳流体系统杰出研究中心 (CCFS) 及地球科学研究所，
科廷大学应用地质系，澳大利亚，珀斯
2 浙江大学地球科学系，中国，杭州
3 中国科学院地质与地球物理研究所，中国，北京

主要英汉译者：孟立丰

考察目的

通过对典型野外露头的观察，认识华南东部的地壳组成及塑造今日华南地块的主要地质事件，并通过此过程提高参加者由野外观察及测试数据来分析解释地质过程的能力。

适合对象

任何具有地质学学位，且对解读岩石记录及了解华南地质历史有兴趣的研究者。

谨言

人们常说我们只能看到已知的或者自以为已知的现象，但忘记了岩石也能向我们叙说我们尚未知晓的知识或故事。

我们将看到的可能是不完整的地质记录，认识也不一定正确，因此每次观察同一个野外露头都应该有新的收获。

面临的问题和挑战

发现新的现象，提出更好的模式，并设法去检验它。

罗盘校正

实习区域罗盘磁偏角在 2013 年 7 月为西偏 4.2°~5.1°，并于每年继续西偏约 0.05°。

(http://www.ngdc.noaa.gov/geomag-web/#declination)

---

[*] 其中一些野外地质点描述改编自 Li et al. (2003b); 图件中所用野外照片除非有特别标注，皆由李正祥提供

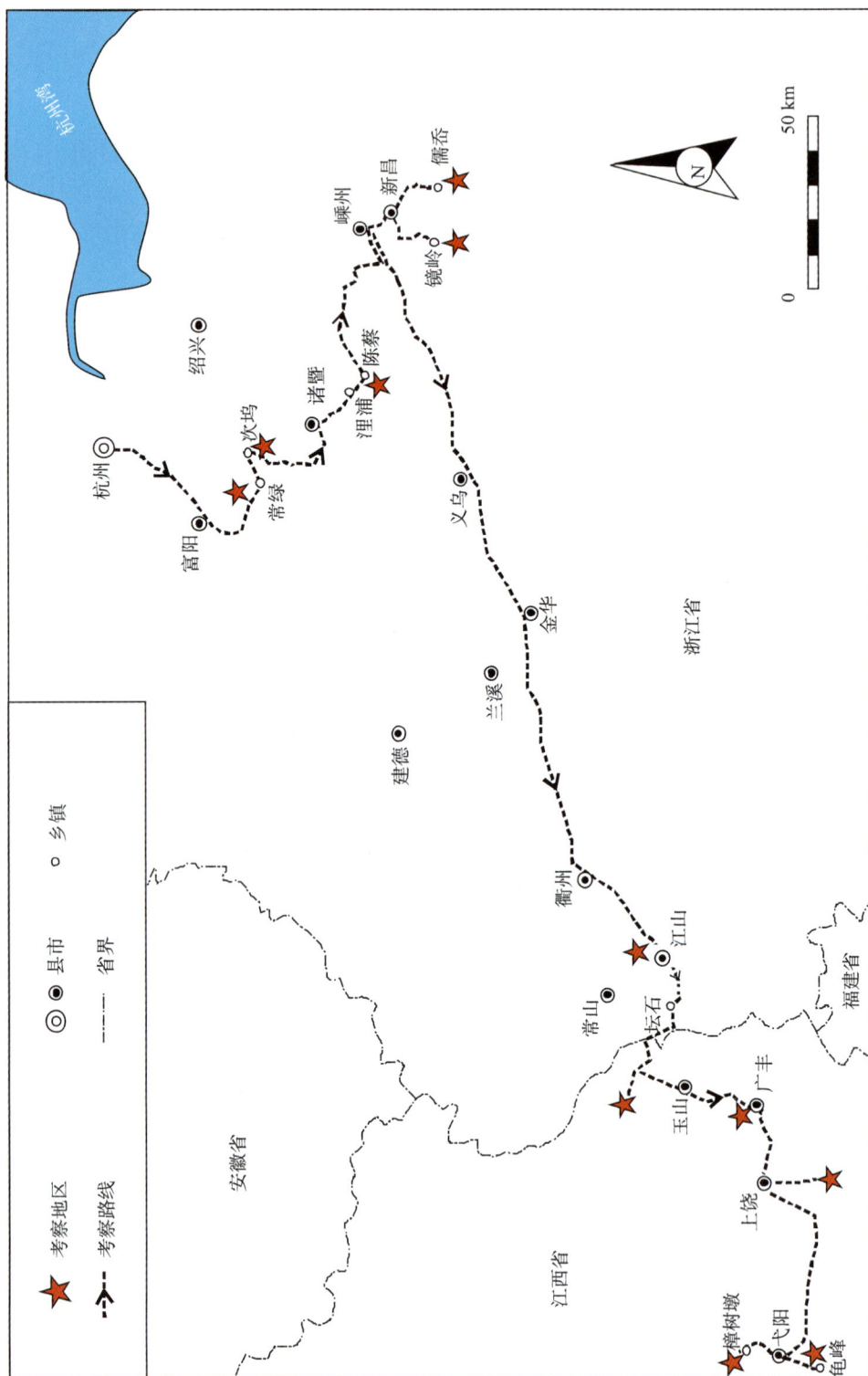

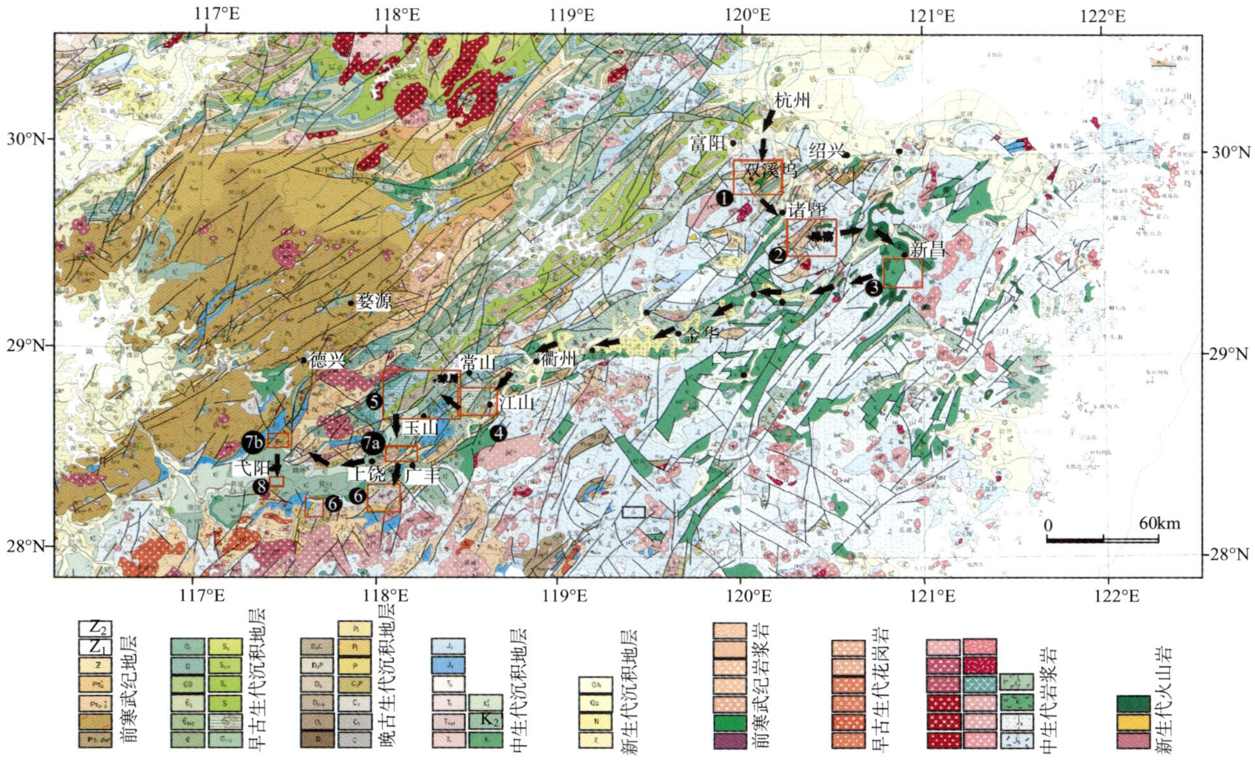

图 2.1 主要地质考察点的位置及相关的地质考察内容。地质图据 1:1 50 万《中国地质图集》

（马丽芳等，2002）中浙江幅、江西幅和福建幅等修编

❶ 从四堡造山作用到新元古代裂谷岩浆作用
❷ 陈蔡杂岩：华夏地块古老基底，早古生代武夷-云开造山
   带及江-绍断裂
❸ 晚中生代大火成岩省（二叠-三叠武夷-云开造山后岩浆作用）
❹ 江山地区下古生界武夷-云开造山后的盆地记录
   及该造山与印支造山事件的构造变形

❺ 下古生界三清-少华地层剖面：关于武夷-云开造山带的
   远端盆地沉积响应记录及下古生界与古生界的平行不整合
● 二叠-三叠印支造山作用到造山后岩浆作用
● 四堡朗田里片岩、樟树墩蛇绿杂岩及新元古代裂谷岩浆作用
● 飞阳龟峰国家地质公园：白垩纪红层与丹霞地貌

第 1 日：从四堡造山作用到新元古代裂谷岩浆作用

考察目的：认识华南地块 (SCB) 新元古代构造体制的转变过程，即从中元古代晚期—新元古代早期的板块聚合造山作用 (即以往称为"神功造山运动"，区域上的"四堡造山运动"或"江南运动") 转变为新元古代中期的构造伸展作用 (即南华裂谷作用)。

考察内容：双溪坞弧火山碎屑岩层序，神功 (四堡) 造山作用形成的角度不整合，裂谷盆地底部层序及 850~780 Ma 的双峰式岩浆侵入体。

早晨从浙江大学玉泉校区出发前往富阳市双溪坞地区。

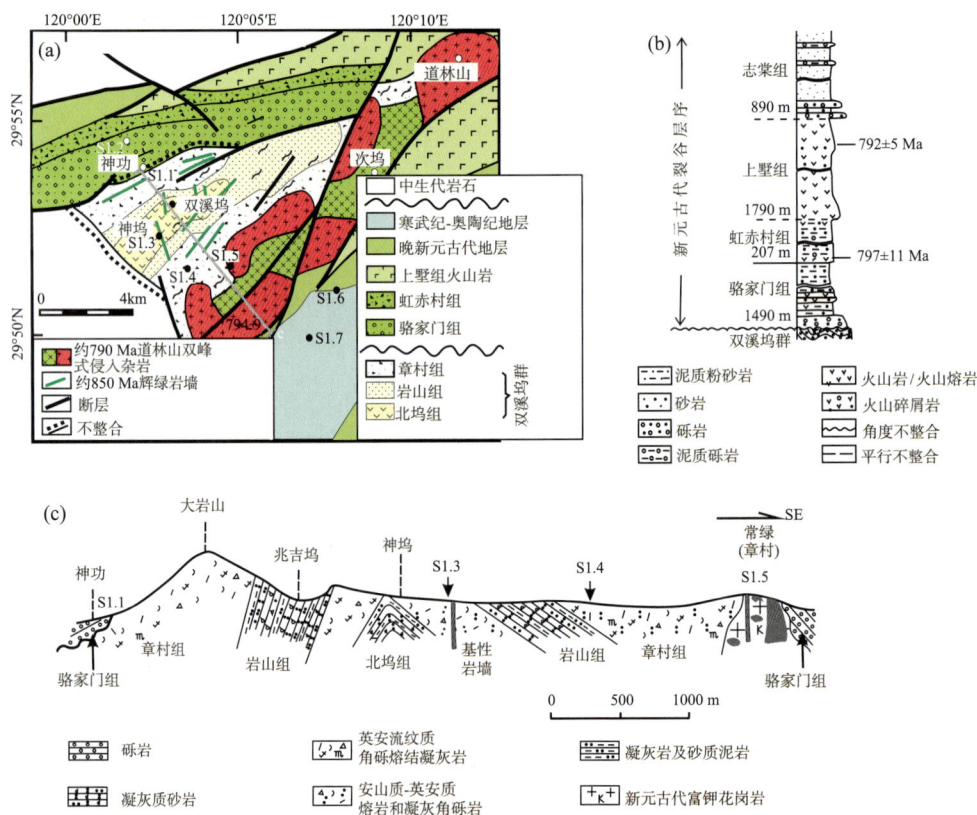

图 2.2 (a) 双溪坞地区地质简图 (Li et al., 2009)；(b) 双溪坞地区新元古界柱状图 (Li et al., 2003a)；(c) 双溪坞背斜剖面图 (据浙江省地质矿产局 (1989) 改编)

➢**考察点 1.1　位于神功村（现名勤功村）的神功不整合 (29°53.302′N，120°02.316′E)**

观察 970~890 Ma 的双溪坞弧体系 (Li et al., 2009) 与上覆的新元古代骆家门组之间的接触关系 (图 2.2)。骆家门组的底砾岩不整合覆盖在岩性成分变化且凹凸不平的古剥蚀面之上。该不整合面是确定四堡造山作用 (前人也称其为神功运动，因其位于神功村而得名) 的关键性地质证据，并被认为是华南板块在早新元古代从构造汇聚转变成陆内裂谷的响应 (图 1.3)。

底砾岩砾石的直径大多为 2 至 10 cm 不等，最大可达到 50 cm 左右，且普遍分选较差，呈次棱角-次圆状。砾石岩性成分复杂，有花岗岩、基底变质岩和火山碎屑岩 (图 2.3)。不整合面之下的章村组凝灰岩被强烈劈理化。章村组层面 (从其成分条带变化判别) 及其劈理面产状都近似直立。值得注意的是，章村组内发育的北东向劈理并没有延伸至上部新元古代骆家门组，而且骆家门组的产状较缓 (在神功村附近倾向为北西，倾角约 20°)。

图 2.3　神功村骆家门组的底砾岩

骆家门组为新元古界河上镇群的底部沉积单元，前人解释为一套大陆裂谷底部层序 (Wang and Li, 2003)。其沉积粒度整体上具有向上逐渐变细的趋势。穿过神功村，我们可以观察到厚约 30 m 的骆家门组下部凝灰质角砾岩露头，往上逐渐过渡为凝灰质砂岩。据未发表的锆石 SHRIMP 定年结果，其中的凝灰质角砾的年龄主要分布在 806 Ma 至 922 Ma 之间，反映了其约 800 Ma 的沉积年龄及与周围岩石密切的继承关系 (Li et al., 2003b)。另外，对其上覆的虹赤村组和上墅组 (图 2.2 (a)，(b)) 双峰式火山岩的 SHRIMP 定年显示其年龄分别为 797±11 Ma 和 792±5 Ma (Li et al., 2003a; Li et al., 2008)。同时在该巨厚的层序中不同部位测得了稳定的 800 Ma 左右的年龄显示其具有快速沉积的特征，这不仅需要有足够的可容空间 (裂谷盆地)，而且需要有充足的碎屑物质供给 (强烈的火山活动提供了火山岩和火山碎屑岩，裂谷盆地周边山肩的快速剥蚀也提供重要物源)。

➤ **考察点 1.2　骆家门组上段和白垩纪岩墙 (29°54.138′N，120°01.631′E)**

在这个短暂考察点，主要观察：①具水平层理的新元古代骆家门组凝灰质层段；②侵入该套地层的白垩纪岩墙。此点位于骆家门组的中上部，以细粒凝灰质岩石为主要岩性特征。而火山灰已蚀变为绿泥石、绿帘石及硅质的集合体。侵入骆家门组的二长花岗斑岩岩墙具有明显的冷缩柱状节理和斑状结构，斑晶主要为斜长石、钾长石和石英。对该岩墙 SHRIMP 锆石的 U-Pb 定年给出 118±3Ma 的 $^{206}Pb/^{238}U$ 的加权平均年龄 (王剑等人未公开发表资料，转引自 Li et al., 2003b)。该年龄结果表明此地区岩墙大部分是晚中生代岩浆事件形成的 (见 1.7 节和第 3 日计划)。

➤ **考察点 1.3　北坞组弧火山岩和 850 Ma 后造山辉绿岩墙 (29°52.120′N，120°02.733′E)**

双溪坞群主要是由强烈变形的火山岩、火山碎屑岩夹长英质凝灰岩、凝灰质砂岩及粉砂岩组成。根据岩性特征自下而上将其分为四个组，即平水组、北坞组、岩山组和章村组 (浙江省地质矿产局,1989)。平水组只在绍兴平水镇附近有出露，主要由蚀变的安山质岩石组成 (图 2.4 (a))。尽管平水组未见与其他三个组有直接接触关系，但它还是被认为是双溪坞群的最早层序。平水组的最小年龄由侵入于平水玄武岩中的桃红岩体给出，其 SHRIMP 锆石 U-Pb 年龄为 913±15 Ma　(Ye et al., 2007)。Chen 等 (2009) 报道了平水组两个火山岩的 LA-ICPMS U-Pb 锆石年龄分析结果，并将它们各自的平均 $^{206}U/^{238}Pb$ 年龄 904±8 Ma 与 906±10 Ma 作为平水组火山岩的形成年龄。值得注意的是，该报道中的 U-Pb 年龄高度不谐和，其 $^{206}Pb/^{238}Pb$ 年龄变化范围为 878 Ma 至 999 Ma。但这两个测试结果中的 $^{207}Pb/^{206}Pb$ 年龄在误差范围内基本一致，给出的年龄平均值为 965±12 Ma，同时这一 Pb/Pb 年龄值与 Sm-Nd 同位素年龄 978±44 Ma (章邦桐等，1990) 相符，所以它应该最能代表平水组火山岩的形成年龄 (Li et al., 2009)。

Li 等 (2009) 报道了双溪坞群火山岩及其相关的英云闪长岩和花岗闪长岩侵入体的年龄普遍在 965±12 Ma 和 891±2 Ma 之间，并认为是一套典型活动大陆边缘的钙碱性岩浆组合 (图 2.4)。

在该考察点，我们将首先考察位于整个双溪坞背斜核部的北坞组，其位于绍兴平水镇以西约 50 km。北坞组厚约 430 m，由安山岩、英安岩、流纹岩和火山碎屑岩夹层组成。其 SHRIMP U-Pb 锆石年龄为 926±15 Ma (Li et al., 2009；图 2.5 (a))。

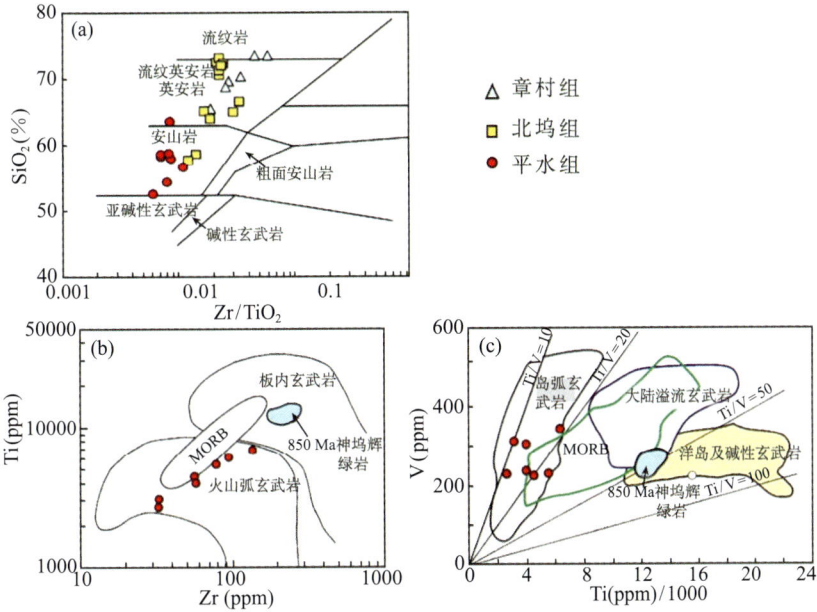

图2.4 (a) 双溪坞群火山岩的岩石分类图，(b) Ti—Zr 相关关系判别图 (Pearce, 1982) 及 (c) 平水组玄武质岩石的 V—Ti 相关关系判别图 (Shervais, 1982)。其中岛弧玄武岩、洋中脊玄武岩 (MORB)、大陆溢流玄武岩及洋岛和碱性玄武岩的范围区域由 Rollinson (1993) 据 Shervais (1982) 绘出 (图引自 Li et al., 2009)

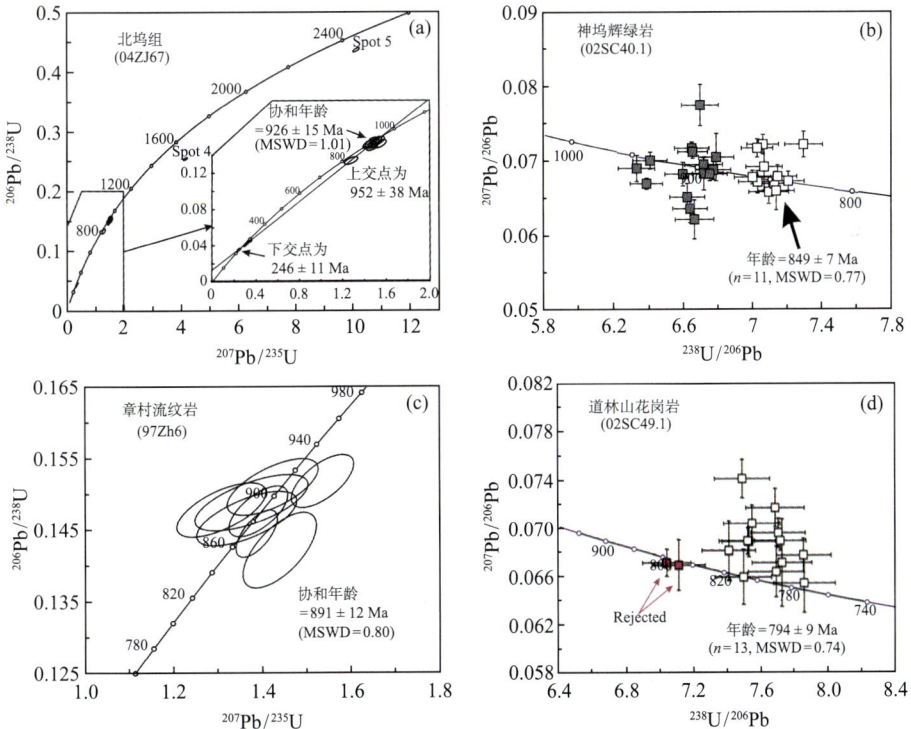

图2.5 双溪坞群各组典型岩石的锆石 SHRIMP U-Pb 年龄结果图：(a) 北坞组流纹岩，(b) 神坞辉绿岩岩墙，(c) 章村组流纹岩及 (d) 道林山花岗岩 (Li et al., 2009)

在考察此点我们还将观察多条弱变形的铁镁质岩墙近垂直侵入双溪坞群弧火山岩。其中的一条岩墙测得的年龄为 849±7 Ma (图 2.5 (b))。同时，这些岩墙的地球化学分析结果显示其具有大陆裂谷型的板内玄武岩特征 (图 2.4 (b) , (c) , 图 2.6 (a) , (b) ; Li et al., 2009)。以上结果表明双溪坞火山岛弧在 890 Ma 左右结束发育，并在 850 Ma 左右从构造挤压环境完全转变为构造伸展环境 (Li et al., 2009)。

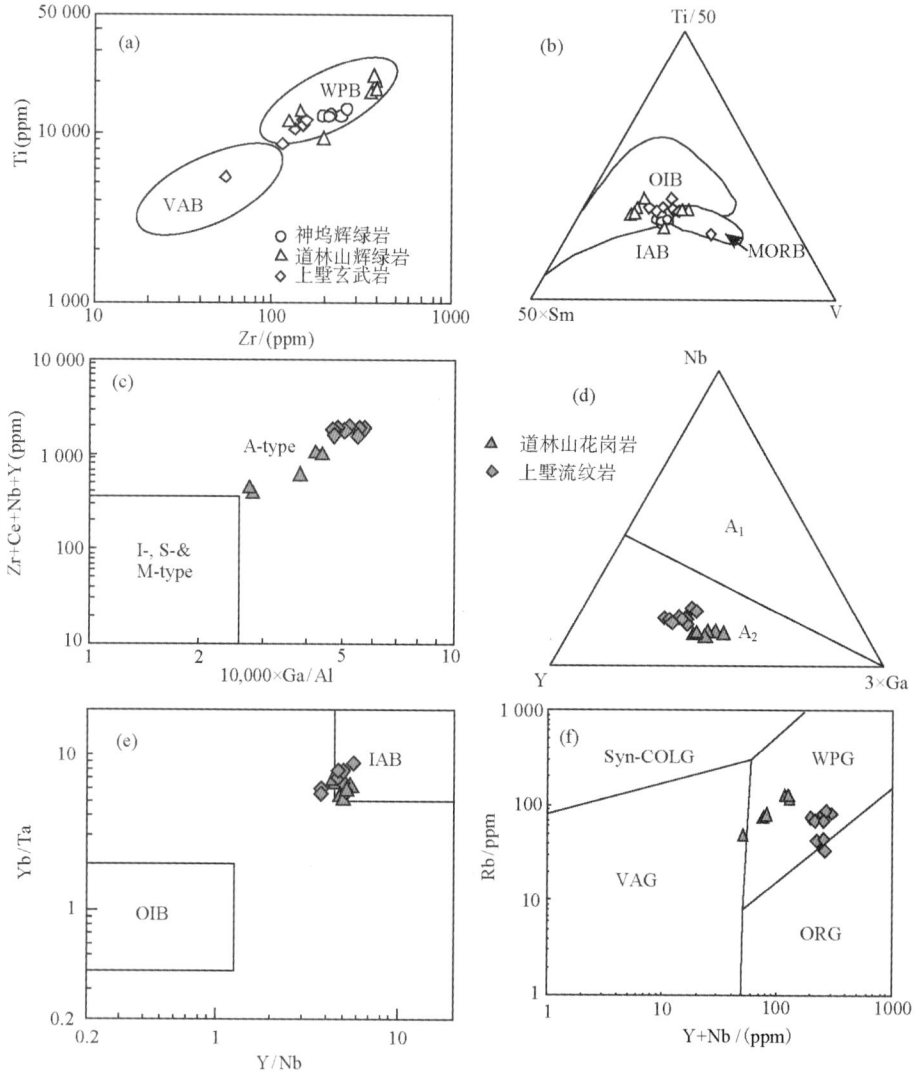

图 2.6 (a) (b) 为浙江北部地区中新元古代镁铁质岩浆岩 Ti-Zr (Pearce, 1996) 及 Ti-Sm-V (Vermeesch, 2006) 地球化学元素判别图。(c)—(f) 为浙江北部地区中新元古代长英质岩浆岩判别图：(c) (Zr+Ce+Nb+Y)—10 000×Ga/Al (改编自 Whalen et al., 1987) 显示其亲 A 型花岗岩特性；(d) Nb-Y-Ga 三元图中 (Eby, 1992) 所有结果都落在 A2 亚型花岗岩区域中；(e) Yb/Ta—Y/Nb 协变图中 (Eby, 1992) 则显示出与岛弧玄武岩的相似特征；(f) Rb—(Y+Nb) 协变关系图中 (Pearce et al., 1984) 则都落在板内花岗岩区域内 (原始图件引自 Li et al., 2008)

➢考察点 **1.4** 章村组熔结凝灰岩——双溪坞弧最后一期岩浆作用 **(29°51.834'N, 120°03.789'E)**

章村组发育于双溪坞群的最上部，主要由约 850 m 厚的长英质熔结凝灰岩组成 (图 2.4 (a))。其锆石的 SHRIMP U-Pb 年龄为 891±12 Ma (图 2.5 (c)，Li et al., 2009)。

此考察点的重要性在于认识章村组为华南最年轻且无争议的新元古代弧火山岩。它可能标志着扬子与华夏地块之间洋盆关闭之前最后一期双溪坞弧岩浆作用事件 (图 1.3、图 1.5、图 1.7 (b))。

➢考察点 **1.5** **ca.790 Ma** 的道林山花岗-辉绿杂岩体——新元古代双峰式岩浆作用的一部分 **(29°51.673′N，120°04.787′E)**

道林山花岗岩-辉绿岩构成的杂岩体主要出露在一个约 5km×30km 的近北东方向区域 (图 2.2)。其主要由粉红色的钾长花岗岩及黑灰色的辉绿岩组成。该杂岩体侵入于四堡期双溪坞群之中。在野外露头上，辉绿岩体大部分与花岗岩呈侵入接触，而在接触带附近可见花岗岩脉出现在辉绿岩体内。花岗岩与辉绿岩体接触边界一般不规则。在一些露头区还发现了辉绿岩与花岗岩侵入体之间存在塑性互动的证据。以上所有观察现象表明铁镁质岩浆与钾长花岗岩至少是同期岩浆作用，在某些地方甚至可能表明它们为同源岩浆作用。其花岗岩样品的锆石 SHRIMP U-Pb 的年龄为 794±9 Ma (图 2.2 (a)，图 2.5 (d))。

Li 等 (2008) 研究发现道林山地区发育的镁铁质岩石具有拉斑玄武岩特征，并认为它与区内的上墅组玄武岩很有可能来源于一个共同的软流圈地幔。而不同壳源组分的同化混染作用在这些玄武质岩石的形成过程中影响很小。他们还发现道林山钾长花岗岩和上墅组流纹岩都具有强烈的铝质 A 型花岗岩的亲缘性元素特征，并认为它们由先存的四堡期弧钙碱性火成岩在低程度脱水熔融作用下并经历不同程度的结晶分异作用形成。这些火成岩均为非造山岩浆作用产物 (图 2.6)。Li 等 (2008) 进一步提出约 850 Ma 的神坞辉绿岩可能代表了非造山岩浆作用的开始，也是南华大规模裂谷作用开始的先兆，而在裂谷发育中期则产生了约 790Ma 的双峰式岩浆活动。Wang 等 (2010) 报道了道林山杂岩体东部 790 Ma 的 A 型花岗岩的存在，将其解释为较老 (850~800 Ma?) 的底侵于下地壳的拉斑玄武质岩石的高温熔融产物，并认为其与 Rodinia 超大陆裂解期间的地幔柱活动相关。

备选点：

如果时间充裕，在下午去诸暨的路上，我们可能沿途增加考察点，以考察路边中新元古代—早古生代地层。值得注意的是，增加的考察点位于双溪坞岛弧与江-绍断裂带

之间 (详见第 2 日计划)。

➢**考察点 1.6** 新元古代裂谷上部层序 (志棠组火山碎屑岩) 与寒武系底部 (荷塘组) 炭质页岩断层接触 **(29°50.968′N，120°7.811′E)**。目前志棠组尚无精确的年代学限制，通过横向对比认为其年龄可能为 **ca. 750 Ma**

➢**考察点 1.7** 中寒武世杨柳岗组：薄-中层的中等程度变形但尚未变质的灰岩 **(29°49.832′49.9N，120°6.988′E)**

当日作业：

- 用数值年龄作为纵轴，绘制双溪坞地区简单的构造地层柱状图。

当日问题思考：

- 如何利用地质和地球化学信息区分岩浆作用是与裂谷或地幔柱作用有关还是与俯冲作用有关？

第 2 日：陈蔡杂岩——华夏地块的古老基底，早古生代武夷-云开造山作用及江绍断裂带

考察目的：我们将考察华夏地块基底代表性岩石的原岩，理解早古生代武夷-云开 (加里东期) 造山作用发生的时间和性质，以及江绍断裂的地质含义—所谓的扬子与华夏地块间拼合界线。

考察内容：陈蔡杂岩岩性——包括富含石榴子石片麻岩、花岗片麻岩 (片麻状花岗岩？)、大理岩和 1.78 Ga 的变质-辉长岩墙/岩床及围岩片麻岩 (该地区发现的最古老岩石)；陈蔡杂岩体的变形样式；840 Ma 的泩浦/璜山辉长-闪长岩；江绍断裂带。

陈蔡杂岩 (习惯上称为"陈蔡群") 位于浙江省中部，主要由片麻岩、角闪岩、石榴石-白云母片岩及大理岩组成，并具有北西向主导的构造推覆方向 (图 2.8 (a) (b))。陈蔡杂岩体的研究一直以来是华南地块构造研究中的难题。由于陈蔡杂岩体经历了中-高级的变质作用，具有复杂的岩石学及构造特征，加上早期可靠年龄数据的缺乏，前人对其存在多种解释。主要观点包括：①其为 > 900 Ma 华夏地块的基底 (水涛，1987；Zhou and Zhu, 1993)；②它是 1400~900 Ma 的蛇绿混杂岩 (Li, 1993)；③其为中-新元古代岛弧 (孔祥生等，1995)。孔祥生等 (1995) 报道了陈蔡杂岩体内部存在大量铁镁质-长英质变质火山岩及侵入体，其中镁铁质岩石可从拉斑玄武岩变化到玄武质安山岩和凝灰岩。该杂岩体北部被江绍断裂 (带) 切断 (图 2.8)，且与已知的古生代沉积地层无直接接触关系，但被早侏罗世或更年轻的地层以角度不整合覆盖 (图 2.7；浙江省地质矿产局,1989)。

最近 Li 等 (2010) 揭示了陈蔡杂岩体的原岩主要为新元古代火山岩及火山碎屑岩 (考察点 2.7)，而在考察点 2.7 的北部不到 10 km 处出露弱变质的双峰式火山岩 (年龄为 838±5 Ma)。此外，在该杂岩体中还发现一小部分古元古代基底被 1781±21 Ma 的辉长岩岩床/岩墙侵入 (考察点 2.3)。陈蔡杂岩体大部分遭受了早古生代 (武夷-云开期) 的变形和变质 (图 2.8 (b))，如角闪岩相变质时间为 ca. 450 Ma (考察点 2.3 和 2.7)，其减压熔融发生时间为 ca. 430 Ma (考察点 2.3 和 2.5)，角闪石冷却年龄为 ca. 425 Ma (图 1.3，图 1.16 至图 1.21 及概论 1.6 节)。杂岩体中大理岩 (考察点 2.6) 的原岩年龄尚不清楚，推测可能为晚新元古代或者早古生代。

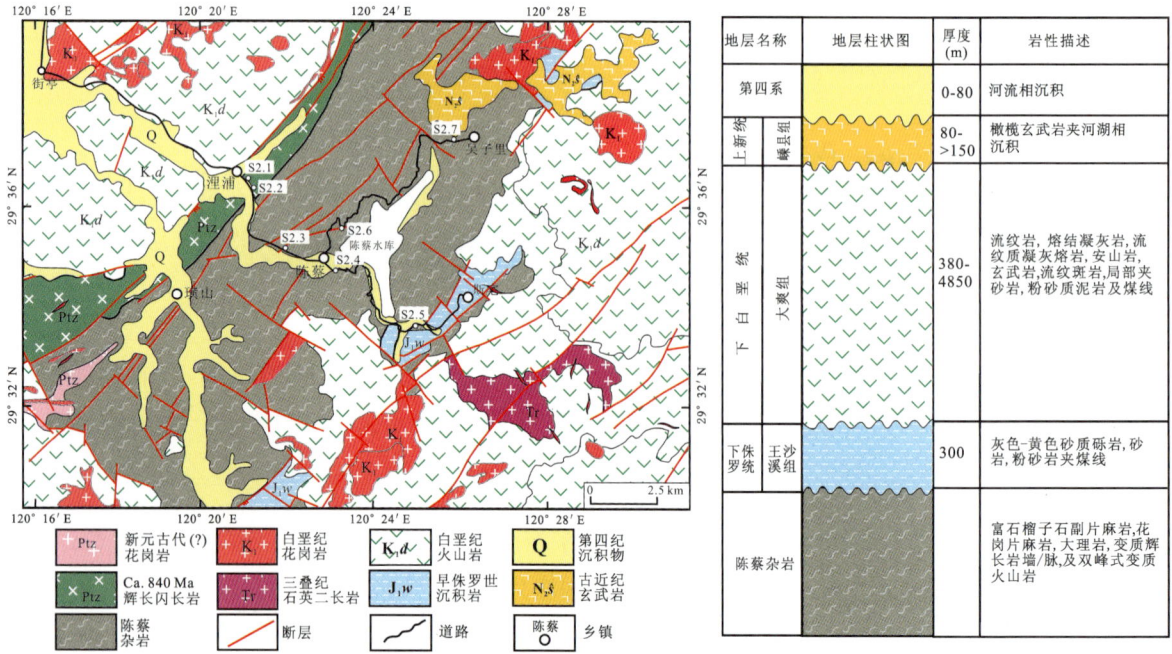

地层名称	地层柱状图	厚度 (m)	岩性描述
第四系		0~80	河流相沉积
上新统 嵊县组		80~ >150	橄榄玄武岩夹河湖相沉积
下 白 垩 统 大爽组		380~ 4850	流纹岩,熔结凝灰岩,流纹质凝灰熔岩,安山岩,玄武岩,流纹斑岩,局部夹砂岩,粉砂质泥岩及煤线
下侏罗统 王沙溪组		300	灰色-黄色砂质砾岩,砂岩,粉砂岩夹煤线
陈蔡杂岩			富石榴子石副片麻岩,花岗片麻岩,大理岩,变质辉长岩墙/脉,及双峰式变质火山岩

图 2.7 陈蔡地区地质图 (据 1:25 万嵊县幅地质图改编)

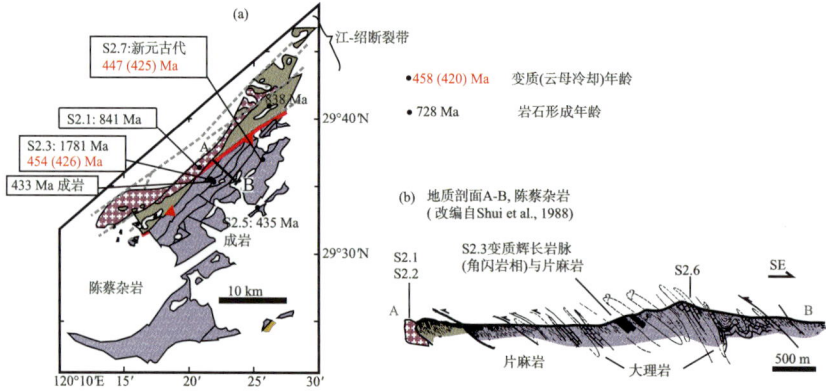

图 2.8 陈蔡杂岩体的年龄 (a) 及其变形构造样式 (b) (Li et al., 2010)

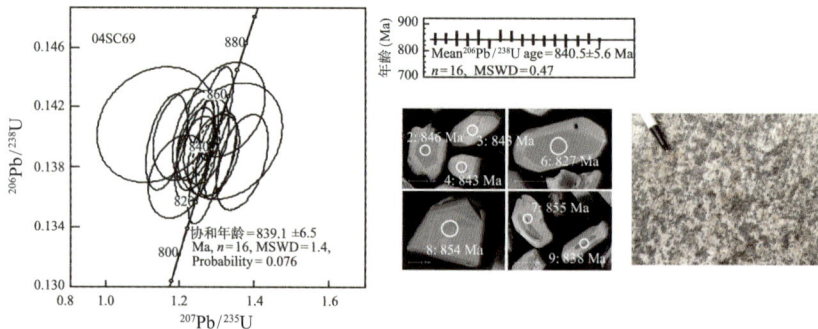

图 2.9 浬浦/璜山辉长-闪长岩的锆石 SHRIMP U-Pb 年龄 (图 2.8 中 S2.1) (Li et al., 2010)

➢ **考察点 2.1　弱变形的 840 Ma 涅浦/璜山辉长-闪长岩 (29°36.627′N，120°20.973′E)**

　　该岩体被数条断裂所切割 (江绍断裂带的组成部分?)，且变形程度不一。目前认为它是与大陆裂谷相关的新元古代中期双峰式岩浆岩的一部分 (概论 1.5 节和第 1 日的考察点 1.5)，但详细的地球化学和岩相学分析工作仍在进行中。

➢ **考察点 2.2　涅浦/璜山辉长-闪长岩中的韧性剪切带 (29°36.377′N，120°21.244′E)**

　　主要观察涅浦/璜山辉长-闪长岩内部北东向近直立的糜棱岩韧性剪切带。作为江绍断裂带的组成部分，该剪切带目前还没有精确的定年数据。

➢ **考察点 2.3　陈蔡杂岩体的古元古代原岩——本区域内最古老的岩石 (29°35.195′N，120°21.882′E)**

　　观察原岩为古元古代 (1781±21 Ma) 的辉长岩侵入体 (岩墙亦或岩席) 变质成为斜长角闪岩。锆石 U-Pb 定年的上交点年龄 1781±21 Ma 被解释为原岩年龄，下交点年龄 454±29 Ma 则是其高级变质年龄 (图 1.19，图 2.10)。该解释与锆石晶体的内部环带以及 Th/U 比变化特征相符 (图 2.10 (a)；Li et al., 2010)。此外，在附近围岩中的混合岩则得到了 2700~2200 Ma 锆石核部年龄及 433±3 Ma 的混合岩化年龄。变质形成的角闪石矿物则给出了 426±1 Ma 的冷却年龄 (图 2.10 (b))。此区域的 P-T 曲线及 P-T-t 轨迹图见图 1.19。

　　注意测量陈蔡杂岩体的主要构造面理产状。

➢ **考察点 2.4　陈蔡变质杂岩体亦或白垩纪花岗岩？(29°34.736′N，120°23.525′E)**

　　该点为短暂考察点，主要为了核实最新 1:25 万区域地质图上标定的沿坝下公路出露的岩石是属于陈蔡变质杂岩还是白垩纪花岗岩 (图 2.7)。

　　注意石榴子石和变质铁镁质岩墙的出现。

　　测量主要的构造面理产状并与点 2.3 进行比较。

➢ **考察点 2.5　陈蔡杂岩体中的花岗片麻岩/片麻状花岗岩 (29°33.739′N，120°25.163′E)**

　　花岗片麻岩/片麻状花岗岩在通往斯宅的公路桥下出露较好。花岗片麻岩中富含石榴子石且很可能为副片麻岩 (图 2.11 之左上图片)。顺着小溪往下，在桥另一侧的岩石更多表现为片麻状花岗结构 (图 2.11 之右上图片)。片麻状花岗岩锆石中只有少量 ca. 800 Ma 和更古老的核部年龄，并且这些锆石具有类似岩浆锆石的环带图 (图 2.11) 及中等的 Th/U 比。而绝大部分锆石给出的 435±4 Ma 年龄 (图 2.11) 则可解释其为同造山—造山晚期花岗岩的重熔作用年龄。值得注意的是，此重熔年龄与考察点 2.3 中混合岩化作用的年龄 433±3 Ma 基本一致（图 2.10 (c)）。

根据最新地质图（1:25 万嵊县幅），该考察点位于白垩纪花岗岩的边部——怀疑是该图填图解释有错误。

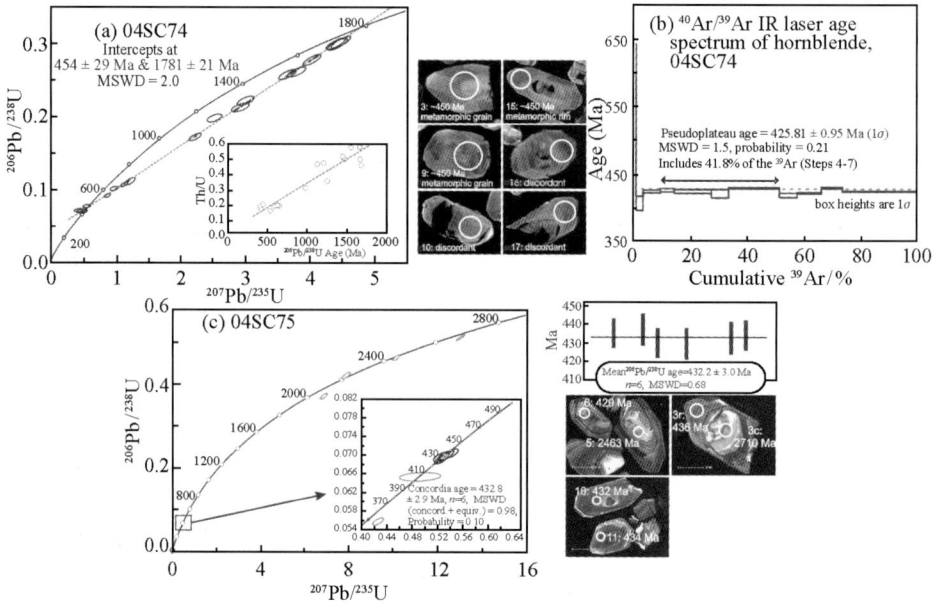

图 2.10 陈蔡杂岩体中变质辉长岩墙（或岩席）的 (a) 锆石 SHRIMP U-Pb 年龄，(b) 角闪石 ^{40}Ar/^{39}Ar 坪年龄，及 (c) 其附近围岩中混合岩样品的锆石 SHRIMP U-Pb 年龄 (Li et al., 2010)

测量主要构造面理产状，并将它与前两个考察点进行比较。

➤考察点 2.6 陈蔡杂岩体中大理岩出露区（29°35.431′N，120°23.088′E）

沿马路边考察陈蔡杂岩中大理岩的露头。在前方马路急转弯处的一个废弃大理石采石场内发现有角闪岩。虽然目前尚无精确的年代学限定，但笔者认为大理岩的原岩可以与闽西北地区的新元古代马面山群中的大理岩作对比。马面山群的岩性包括大理岩、绿片岩及变质火山岩，其中变火山岩的年龄为 818±9 Ma（Li et al., 2005）。然而不确定的是，这些大理岩的原岩年龄也可能很年轻。

➤考察点 2.7 吴子里村陈蔡杂岩中富石榴子石副片麻岩（29°37.360′N，120°26.002′E）

在石桥下的小溪床内，可以观察到富石榴子石副片麻岩（图 2.12, 样品 01SC40）。样品 04SC88 是取自附近的与 01SC40 类似的富石榴子石片麻岩，其黑云母 ^{40}Ar/^{39}Ar

年龄从核部到边部具有系统的变化特征。其最大核部年龄为 425±4 Ma。此 425±4 Ma 黑云母核部年龄与陈蔡杂岩体样品 04SC74 给出的角闪石冷却 (约 500°C) 年龄 426±1 Ma 在误差范围内一致，因此可以认为陈蔡杂岩体从 500°C 冷却至 300°C 的速度非常快 (Li et al., 2010)。

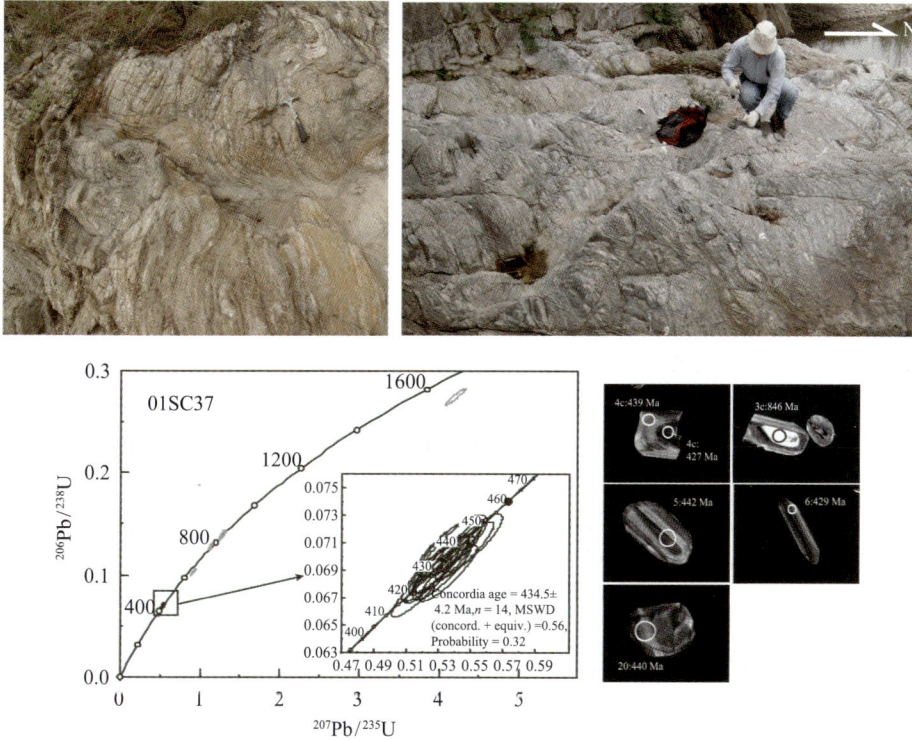

图 2.11　斯宅地区片麻状花岗岩的野外照片及其锆石 SHRIMP U–Pb 年龄 (Li et al., 2010)

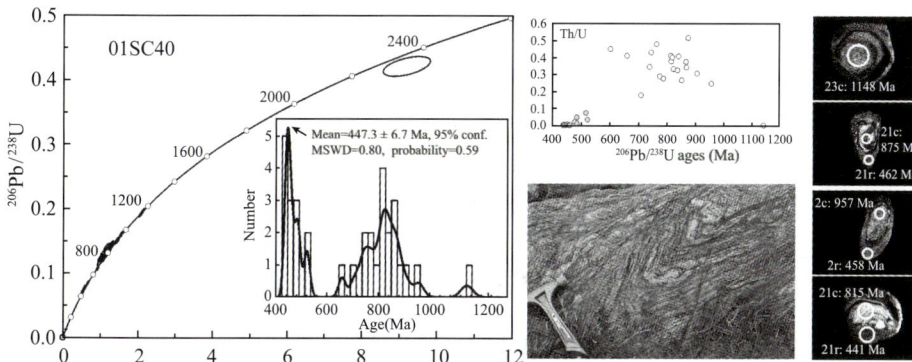

图 2.12　吴子里陈蔡富石榴子石副片麻岩的锆石 SHRIMP U-Pb 年龄 (Li et al., 2010)

当日作业:

- 绘制陈蔡地区简单的构造地层柱状图,并将其与双溪坞地区相比较 (所有柱状图使用相同的形式及垂向比例尺) 。

当日问题思考:

- 扬子和华夏地块的边界在何处?在双溪坞弧及江绍断裂带之间应该会有什么基底:是扬子地块还是华夏地块的基底,还是两者都有可能?
- 为什么江绍断裂带南部没有出现任何古生代地层?

雪后的皇渡桥 (S3.3) (李正祥摄)

第3日：晚中生代 (二叠纪–三叠纪印支造山作用之后) 大面积岩浆作用

考察目的：考察华南东部晚中生代大面积岩浆活动产出的白垩纪双峰式火山岩和侵入岩，并讨论它们的构造意义。

考察内容：镜岭白垩纪双峰式火山岩层序 (令人激动的火山构造)，儒岙双峰式侵入杂岩体，白垩纪硅化木及其围岩。

侏罗纪-白垩纪岩浆作用广泛分布于华南中东部地区，它们的构造意义至今仍存在争论 (概论 1.7 节和图 1.25)。尽管浙东地区中生代岩体在时间上要追溯得更远，但是覆盖大部分地区的火山岩时代却以白垩纪为主。此区域发育了一个世界级的古火山群，其中一些好的火山构造和地貌景观被建成了地质公园而加以保护。但据我们所知，至今仍未有一本专门描述该地区壮观的火山构造和古地理的著作。

➤考察点 3.1 新昌县镜岭镇早白垩世双峰式火山岩层序 (29°21.785′N，120°46.642′E)

我们将在镜岭镇南 (图 2.13) 从山脚出发走向山脊顶部的隘口，沿途观察早白垩世馆头组双峰式火山岩和火山构造 (图 2.14)。

➤考察点 3.2 新昌澄潭硅化木地质公园 (29°23.713′N, 120°46.850′E)

我们将在这里观察若干早白垩世地层中保存的巨型石化树干 (图 2.15 (a))，并分析可能的硅化木形成过程。在该剖面中发现了 6 层地层产有硅化木，并且都集中在馆头组的中段。无论有无根部保留，大部分巨型树干化石近于与层面平行平躺。另一些虽然只保留树桩，但仍保持直立状态。经过利用锆石 SHRIMP U-Pb 年代学方法对其中第三层硅化木层的凝灰岩样品进行分析，得到其年龄约为 119~116 Ma (章凤奇等，未发表数据)。

前人关于本地区硅化木成因的一个说法认为是火山活动驱动热流体通过研究区内老的富硅砂岩层形成硅质热流体，导致已死掉的树木被硅化。但有一个有意思的现象是，

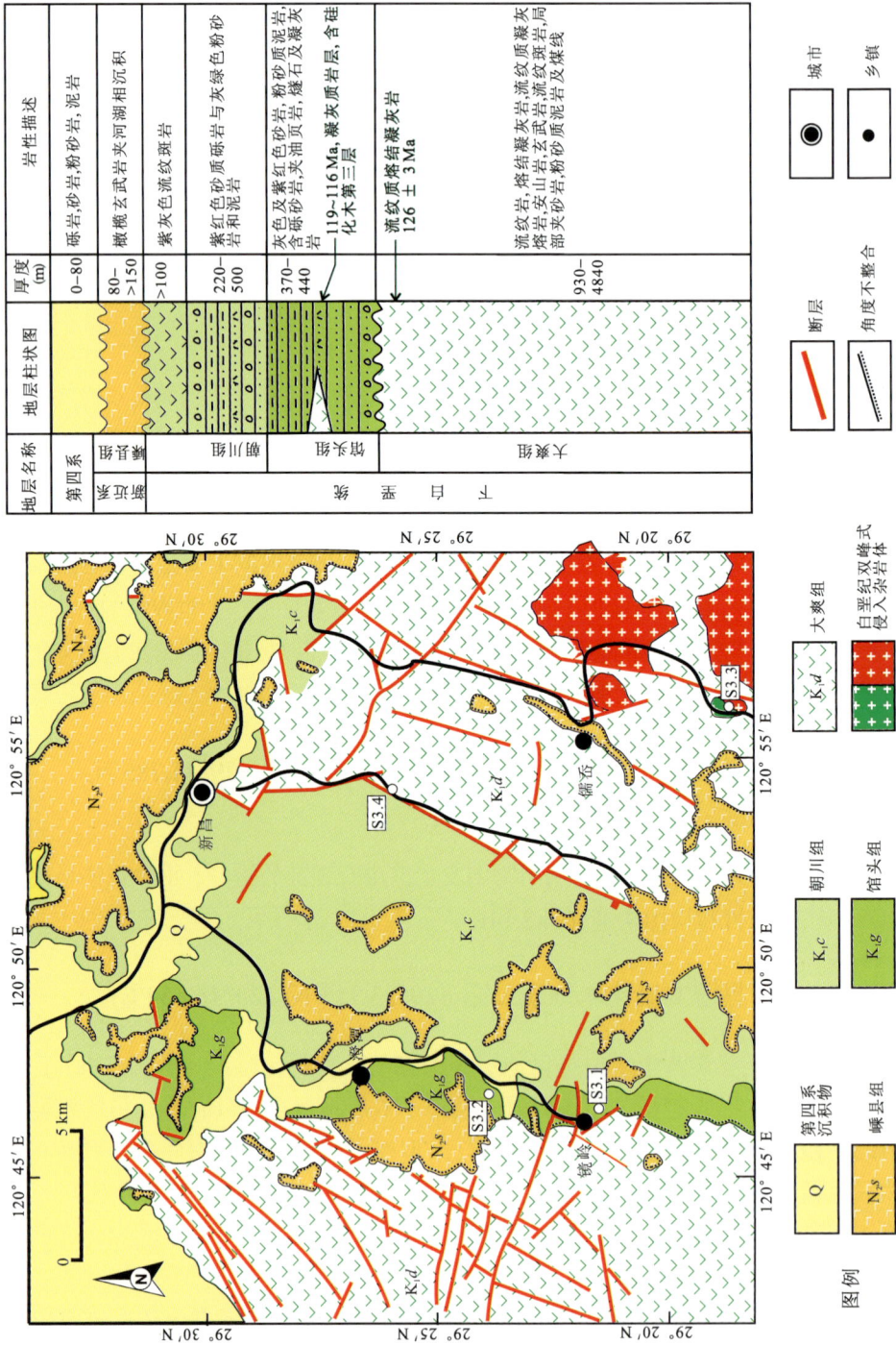

图 2.13 新昌地区地质图（据 1:25 万嵊县幅和 1:20 万诸暨幅地质图修编）

图 2.14 (a) 固结的玄武质熔岩流形成了核部为致密玄武岩, 外部被杏仁状玄武质外壳包围的结构; (b)、(c) 可能为由两期熔岩流形成的一熔岩管; (d) 管内被火山角砾填充 (Li Z X et al., 本次成果)

图 2.15　(a) 澄潭地质公园中的树干化石之一；(b)、(c) 1991 年南安第斯 (Andes) 的哈得逊 (Hudson) 火山爆发，喷出的火山灰 ((b) 中的地面覆盖物) 使得距其约 50 km 以外的山谷底部的森林大面积死亡；(d) 澄潭地质公园中硅化木层的地层层位

这些硅化木通常被埋在火山凝灰岩层中。李正祥等 (未公开发表) 推断可能是灼热的火山灰以及被加热了的地表及地下水使得树木凋亡 (图 2.15 (b)，(c) 所示)，而火山灰中滤出的二氧化硅再将这些死亡的树木石化 (该成因的一个证据是在点 3.1 中的凝灰岩及熔结凝灰岩中出现了大量的玛瑙)。

此外，作为一个重要的古海拔的指示，硅化树木的存在也指示了区域地貌低洼处至少位于林木生长线以下 (该线在现今约为 1500 m)。

➤考察点 3.3　早白垩世儒岙双峰式侵入杂岩 (29°18.750′N，120°56.367′E)

我们在这里观察花岗岩和辉绿岩之间的岩浆机械混合作用 (图 2.16)。往西在河床中我们还会看到辉绿岩侵入体，而往东则是儒岙花岗岩体 (在桥的两头露头良好)，其锆石 SHRIMP U-Pb 的年龄为 116±3 Ma (董传万等，2008)。

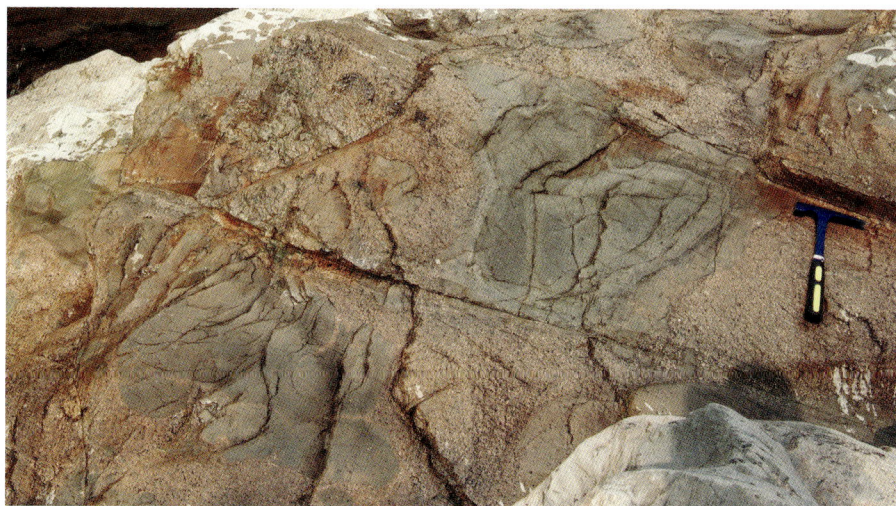

图 2.16　新昌儒岙杂岩体中的岩浆机械混合作用

如果时间允许，我们还可以驱车前往花岗岩岩体中部观察铁镁质岩墙的侵入现象。从更大区域上来看，白垩纪的火山岩及侵入岩都具有双峰式的组分特征。尽管区域内也发现有中性安山岩，但其在晚中生代岩浆岩中只占很小比例。问题是什么大地构造环境才能造成这样大规模的岩浆活动？目前比较流行的构造模型包括：①正常角度俯冲或低角度俯冲的安第斯型大陆弧模型 (Jahn et al., 1990; Zhou and Li, 2000; 这一模型无法解释双峰式岩浆的出现)；②类似于盆-岭式的拉张模型 (Gilder et al., 1991)；③由先前水平俯冲的大洋板片的拆沉导致的盆-岭式拉张，或许叠加有由重新启动的正常角度俯冲而产生的新的大陆岛弧 (Li and Li, 2007; Li et al., 2012; 见概论第 1.7 节)。

Tectonics of the South China Block

➤考察点 **3.4　新昌南部八里村附近早白垩世玄武岩柱状节理 (29°25.937'N，120°54.328'E)**

在这里我们可以看到早白垩世大爽组底部玄武岩发育良好的柱状节理 (图 2.13，图 2.17) 。该节理是在岩浆冷却固结过程中经历热收缩形成。为容纳其收缩量，多边形节理多以近垂直于冷却等温面 (如岩浆流的层面，岩床的底面或者顶面，以及岩墙的墙面——见考察点 1.2) 的产状发育。

图 2.17　新昌南部八里村附近早白垩世玄武岩发育的柱状冷却节理

当日问题思考：

* 你能想到目前世界上还有哪些地方具有与华南东部沿海晚中生代相似的构造环境及地貌特征？

白垩纪地层中形成的丹霞地貌（李正祥摄）

第4日：江山地区的早古生代地层——武夷-云开造山作用开始及印支造山作用的构造印迹

考察目的：早古生代地层，以及这些地层与介于奥陶纪碳酸盐岩和石炭纪底砾岩之间的的角度不整合是如何共同记录武夷-云开造山运动的开始及其在此区域的实际影响。

考察内容：沿着近南北向江山—大陈公路分布的寒武纪碳酸盐岩、奥陶纪泥质灰岩、泥页岩，及其与下石炭统长石砂岩间的角度不整合。此路线还横穿了一系列北东向褶皱。

上午从新昌驱车至江山。

考察区位于江山-绍兴 (江绍) 断裂带的北部，以出露晚前寒武纪至古生代连续沉积地层及北东向的构造为特征 (图 2.18 (a)，(b))。Xiao 和 He (2005) 在考察区进行了系统的构造解析，并将区域内广泛发育的向北西方向逆冲的断层解释为扬子地块 (或者下扬子陆块；详见 Xiao 和 He (2005) 中的图 17) 和华夏地块在中生代缝合拼贴的结果，从而支持了 Hsü 等 (1988) 提出的构造演化模式。然而，江山地区发育有早古生代地层与石炭纪地层之间的角度不整合 (图 2.18 (a) 和点 4.4)，证明这一地区 (相当于 Xiao 和 He (2005) 研究区的东南部及中部变形区；图 2.18 (a)，(b)) 的前石炭纪地层中存在早古生代 (武夷-云开期) 的变形事件 (图 1.16，图 1.21)。换言之，我们在这一地区看到的构造应该是武夷-云开期构造与印支期构造叠加的结果。Li 和 Li (2007) 用水平俯冲 (平板俯冲) 模式来解释中生代 (印支期) 的逆冲作用 (图 1.25，概论第 1.7 节)。

考察点 4.1 至 4.4 位于 Xiao 和 He (2005) 划出的东南构造带，而点 4.5 及 5.1 位于其中部带。

> **考察点 4.1** **下寒武统大陈岭组块状白云质灰岩 (28°49.028'N，118°35.708'E)**

此野外露头位于大陈乡南部公路的东侧，块状白云质灰岩以高角度向北倾 (图 2.19)。

> **考察点 4.2** **中寒武统杨柳岗组暗灰色泥质灰岩、厚层白云岩和条带状灰岩 (28°48.627'N，118°35.722'E)**

这里的中寒武世地层仍然以台地碳酸盐岩为主。观察岩性特征 (图 2.20) 及测量层面产状。

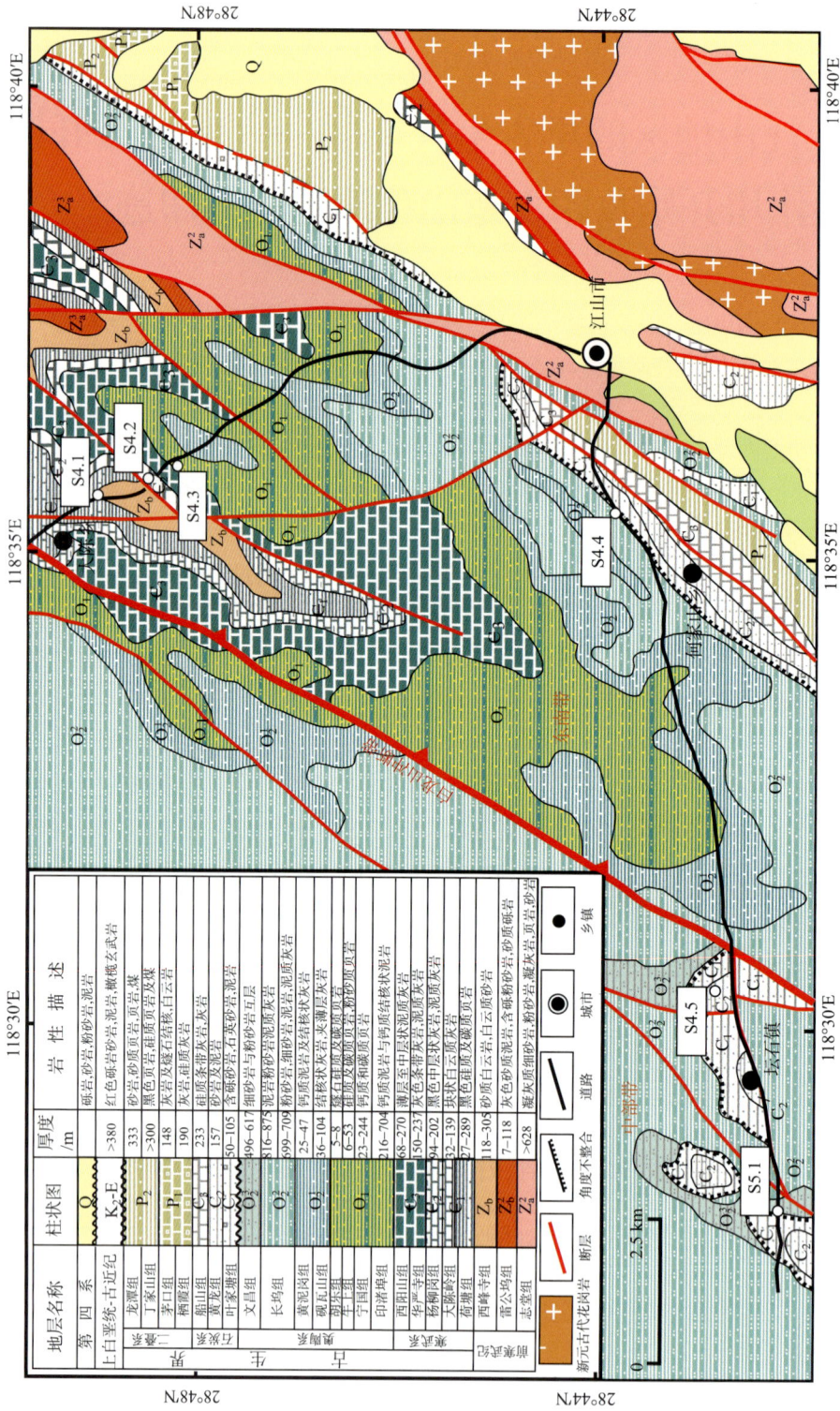

图 2.18 (a)　江山地区区域地质图（据 1:20 万区域地质图），标注有第 4 日的野外考察点及第 5 日的首个考察点位置。构造区带的命名（如中部区带和东南区带）均依据 Xiao 和 He (2005)。注意，在中部区带及东南区带中前石炭纪地层（受武夷-云开事件影响）中的变形历史与西北区带中同年代地层的变形历史具有明显的差别（图 2.24）

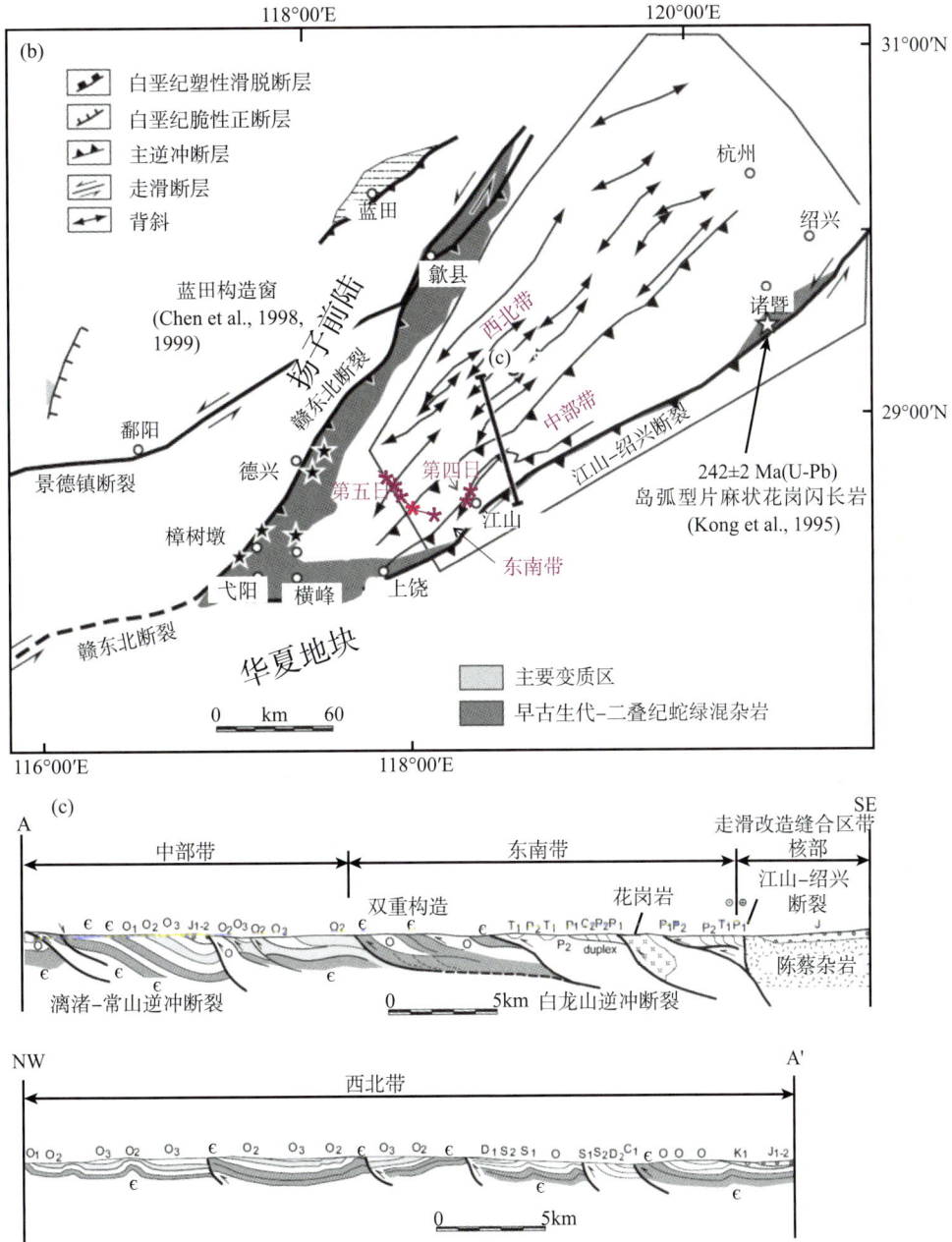

图 2.18 (b)，(c) 浙西北地区构造图及剖面图，展示前陆褶皱冲断带内的构造样式及构造区带。剖面端点 A 及 A'在走向北西—南东的剖面的位置大致一致。K₁-下白垩统；J₁₋₂-下-中侏罗统；J-侏罗系；T₁-下三叠统；P₁-下二叠统；P₂-上二叠统；C₂-中石炭统；D₂-中泥盆统；D₁-下泥盆统；S₂-上志留统；S₁-下志留统；O₃-上奥陶统；O₂-中奥陶统；O₁-下奥陶统；ε-寒武系。空白区代表前寒武基底 (据 Xiao and He，2005 改编)

图 2.19 下寒武统块状白云质灰岩

图 2.20 中寒武统杨柳岗组白云岩 (左) 及薄层状灰岩 (右)

➤考察点 4.3 下奥陶统钙质瘤泥岩及泥岩 (28°48.107'N，118°36.112'E)

该点是区域上首次出现沉积相由台地碳酸盐岩相转变至碎屑岩相。此沉积相转变可能标志着此时的沉积环境由碳酸盐岩台地转变为远源前陆盆地。

我们将从下奥陶统的底部开始观察，关注钙质瘤泥岩的特征 (图 2.21 (a))。而后沿路观察下奥陶统的层序，其中还可以看到褶皱并发育劈理的结核状泥岩 (轴向劈理？) 及褶皱的泥岩 (图 2.21 (d))。从中测量一些层面及劈理面产状，随着测量数据的增多，我们将能够认识区域构造方向，分析构造的形成过程以及研究区在造山带演化过程中的大地构造位置。

➢考察点 4.4 早石炭世与中奥陶世地层的角度不整合 (28°43.93′N，118°35.498′E)

下石炭统叶家塘组长石砂岩与其下中奥陶统的角度不整合接触关系反映此区受武夷-云开造山作用的影响。注意角度不整合面两侧岩层层面产状的不同。

测量奥陶纪地层的层面产状。

图 2.21　早奥陶世钙质瘤泥岩 (a)～(c) 及泥岩 (d)

此处野外露头说明我们沿着横切面 (考察点 4.1 至 4.4 及点 5.1) 所观察到的早古生代地层的变形很可能以武夷-云开 (奥陶纪-志留纪) 期的变形作用为主，而此处相对平缓的石炭纪地层层面产状则显示印支造山的改造作用影响较小。

现在回过头来看图 2.18 (a)，你会注意到尽管该地区的晚古生代地层的构造线与前石炭纪地层的构造线方向近似平行，但是前石炭纪地层的褶皱紧闭程度比其后地层要高得多 (在图 2.18 (a) 左下角尤为明显)。在考察点 5.1，我们也会观察到类似的地层特征关系。此现象与本区印支期变形以薄皮构造样式为特征的解释相符 (图 2.18 (c))。

此考察点还有一个有意思的观察现象是中奥陶世上部层序中的"泥裂"和印模构造 (图 2.22 (a)，(b))，且其岩性由块状碳酸盐岩迅速转变为泥岩。

➢考察点 4.5　中石炭统内部的印支造山期变形 (28°43.041′N，118°30.478′E)

在横渡到坛石公路北侧的一大型灰岩采石场内，中石炭统黄龙组内部箱式褶皱的一翼发育有一良好的逆冲断层 (图 2.23)，反映了"印支运动" (应该称为华南造山作用？详见 Li 和 Li (2007) 的图 1 中的约 1300 km 的华南褶皱/造山带及本指导书图 1.25 中的主冲断层) 的变形作用。如果第 4 天的时间紧张，我们可在随后的第二天早上观察此现象。

图 2.22　彭里西部武夷-云开造山作用的不整合面

图 2.23　江山西北部坛石镇附近的中石炭统内部的印支期逆冲构造

当日作业：

- 绘制一简单的区域构造地层柱状图，形式及比例尺参照第 1 日及第 2 日的作业。
- 利用当日测量的层面产状数据求出下古生界的褶皱轴面和枢纽的产状。

当日问题思考：

- 如何区分该地区的早古生代 (武夷-云开造山作用) 变形与 "印支造山作用" 的变形？
- 武夷-云开造山作用结束于 ca. 420 Ma (详见概论第 1.6 节)。为什么区内缺失志留纪地层？

建造在白垩纪红色块状砂岩与凝灰岩层之上的新昌大佛寺
（The Big Budda Temple）（李正祥摄）

第5日：三清-少华剖面的早古生代地层及漓渚-常山逆冲带两侧的变形差异——武夷-云开造山作用的盆地响应及造山带的分布范围

考察目的： 观察该地区的早古生代沉积环境的变化，即由于武夷-云开造山作用的影响，从碳酸盐台地相沉积转变为前陆盆地相 (粒度向上变粗)，然后从造山后的碎屑岩沉积又变回碳酸盐台地相沉积；分析武夷-云开造山带的北界，并考证同一次造山事件如何在大陆不同地区呈现不同的地质记录。

考察内容： 横穿漓渚-常山逆冲带 (边界断层位于中部及西北部构造带之间)并比较其两侧变形样式的变化；考察三清-少华剖面中的寒武-奥陶纪碳酸盐岩、奥陶-石炭纪碎屑岩和沉积构造、早志留世-晚泥盆世地层之间的平行不整合，以及石炭纪碳酸盐岩。

三清-少华剖面位于 Xiao 和 He (2005) (图 2.18 (b)、(c))中的西北部构造带内，并以发育一系列类似薄皮的北西向冲断岩席为特征 (图 2.18 (b)、 (c), 图 2.24)。我们已经看到在漓渚-常山逆冲带东南方向的东南部及中部构造带内以武夷-云开期 (早古生代)的变形占主导。那么西北构造带内的变形样式是怎样的呢？

> **考察点 5.1 耕读村东部奥陶系中早于石炭纪的强烈变形记录 (28°42.466′N，118°28.094′E)**

此考察点位于 Xiao 和 He (2005)中提到的中部构造带内。在地质图中 (图 2.18 (a))也可以看到强烈变形的奥陶纪地层与轻微褶皱的石炭纪地层之间有一个明显的角度不整合。在此考察点我们主要观察包括：①晚奥陶世文昌组泥质岩发育近垂直的劈理面—试着确定层面并测量层面及劈理面的产状；②早石炭世叶家塘组含砾砂岩的层面产状相对平缓且无劈理发育 (局部受断层影响变陡)。两个组之间的接触界线被覆盖。沿着公路往西则可以看见石炭纪地层疑似被小断层改造后形成的层面产状变化。该考察点上、下地层之间的不整合接触关系主要表现为构造变形样式和沉积环境的显著差异以及长时间的沉积间断，显示了早古生代武夷-云开造山作用对本区的重要影响。

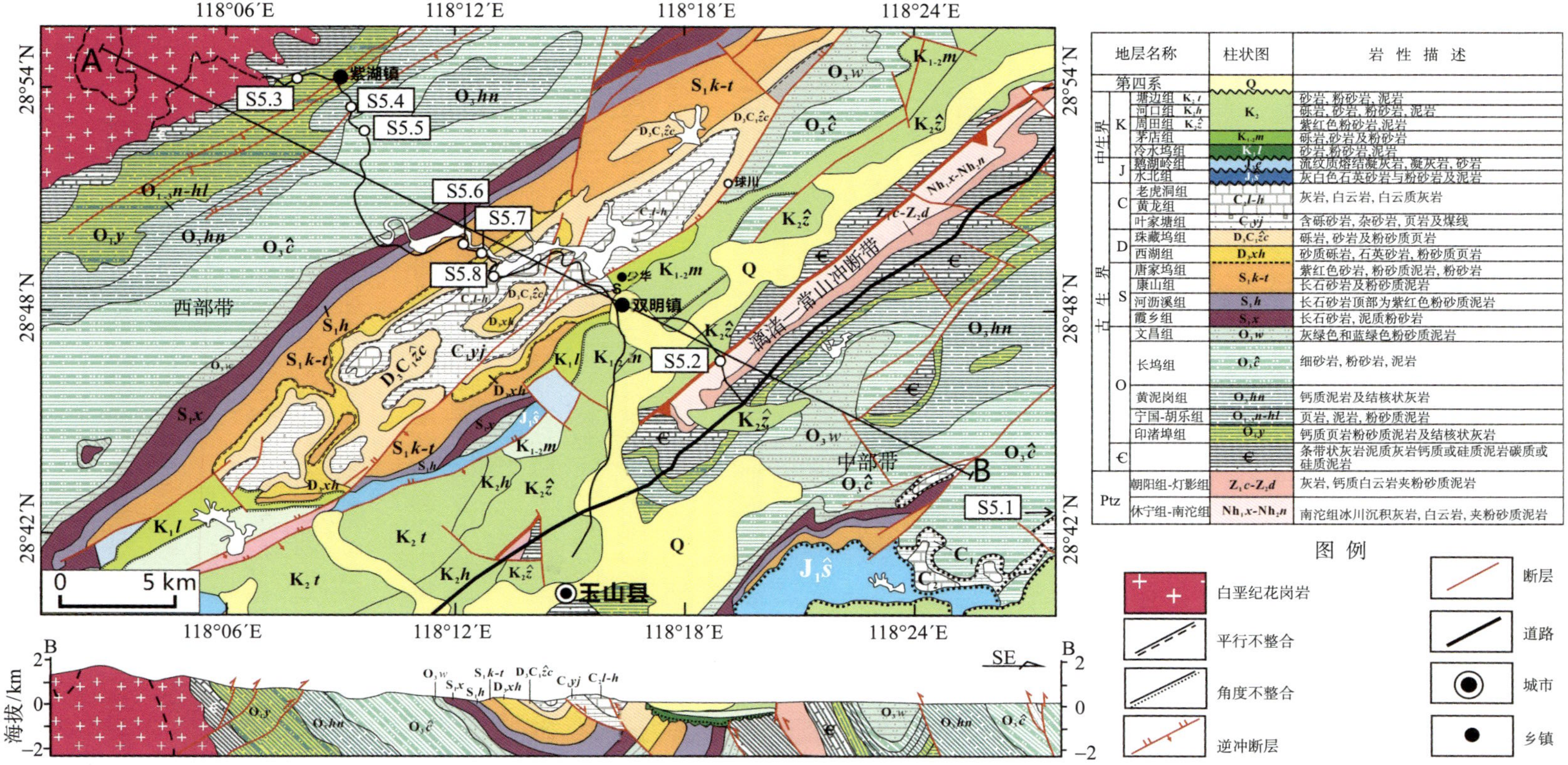

图 2.24 玉山地区地质图及第 5 日考察点 (据 1:25 万上饶幅地质图改编)

注意事项：我们将沿着一条繁忙的公路考察——在你下车后要特别注意安全！

> **考察点 5.2 漓渚-常山逆冲断裂带南侧新元古地层内的次生逆冲构造 (28°46.675′N，118°19.089′E)**

漓渚-常山逆断层是 Xiao 和 He (2005)的构造图 (图 2.18 (b)和图 2.24)中西北部与中部构造带的边界断层。此主冲断层的露头很难找到，但在这里我们可以看到逆冲断层上盘的倾向东南且陡峭的新元古代碎屑岩中发育小型的冲断系统。该冲断系统显示其上盘向北推覆，与主断层的运动方向相似。我们将讨论此逆冲断层的可能活动时间。

在此点短暂停留之后，我们将穿过漓渚-常山冲断带进入西北部构造带。

> **考察点 5.3 寒武—奥陶纪碳酸盐岩 (28°54.123′N，118°07.609′E)**

该考察点再往北仍有寒武纪碳酸盐岩露头，但出露不是很好。本考察点出露硅质/碳质条带灰岩 (图 2.25 (a)，(b))。此地层单元在 1:20 万地质填图中被划为晚寒武世华严寺组，而在新的 1: 25 万地质图中则划为早奥陶世印渚埠组 (图 2.24；此点的岩性特征似乎更符合区域内寒武纪地层的特征，这需要古生物调查及区域地层对比来进一步确认)。

> **考察点 5.4 中奥陶世瘤状灰岩 (28°53.407′N，118°09.202′E)**

将车停靠于公路西侧的东西向桥头后，我们沿着河床往上游方向走。这里可以看到中奥陶统 (黄泥岗组瘤状灰岩露头 (图 2.25 (c))，及发育良好的北东走向的向近直立劈理 (图 2.25 (d))：思考一下这些劈理的可能发育时间。河床两边存在大量非常漂亮的鹅卵石，可根据已看到的地层分析其原岩的可能层位。

> **考察点 5.5 晚奥陶世长坞组细粒砂岩和小型的逆冲断层 (28°53.407′N，118°09.202′E) (将车停靠于考察点南侧的酒店停车场内)**

此点主要观察晚奥陶世长坞组的厚层块状细粒砂岩和粉砂岩，以及波痕、交错纹理和交错层理等沉积构造 (图 2.26 (a))。此层序中厚度巨大的碎屑岩及其浅水沉积环境指示此地区当时既有稳定增长的沉积空间又有丰富的沉积物质来源。这一特征与前陆盆地的构造背景相符。

在此露头上发育有一小型的南东倾向逆冲断层 (图 2.26 (b))——我们可以根据该地区的构造地层记录来讨论此断层可能的形成时代。区域构造特征见图 2.18 (b)、(c)和图 2.24。

图 2.25　寒武纪条带灰岩 ((a) (b)；考察点 5.3)和中奥陶统结核状泥质灰岩 ((c) (d)；考察点 5.4)

图 2.26　晚奥陶世长坞组中的沉积构造及小型逆冲断层

➤考察点 5.6　志留纪紫红色砂岩——前陆盆地末期沉积 (28°49.920′N，118°12.430′E)

我们将考察奥陶-志留纪前陆盆地的最上部的沉积地层单元，即地质图中的 S_1 (早志留世地层) (图 2.24)。此前陆盆地沉积呈现出比下部沉积更粗的粒度，且沉积物颜色变成紫红色,可能反映了随着造山带前锋的靠近,河流相环境的影响持续增加。

> **考察点 5.7 志留纪与泥盆纪地层间的假整合——武夷-云开造山事件的记录 (28°49.657′N,118°12.943′E)**

我们将沿着剖面步行观察造山后沉积层序由老到新的变化特征。假整合面之下的沉积物呈淡红色,可能反映了前陆盆地晚期的河流相沉积环境 (图 1.19)。武夷-云开造山作用结束后,陆块的大部分地区出露地表并持续遭受剥蚀,直到泥盆-石炭纪海侵开始才重新接受沉积,其沉积特征为从底部的底砾岩和石英砂岩 (此考察点的泥盆纪沉积物)开始到最终变为碳酸盐岩台地相沉积 (下一个考察点;图 1.22 (c)、(d))。

在这个考察点需要完成的工作:顺着公路边的露头剖面,看看你能否通过逐层的排查找到志留系与泥盆系之间的假整合——每人都要亲自动手挖掘。在此工作进行过程中需要记录岩性及结构变化,并沿路观察测量层面产状。要求每人根据记录完成一幅简单的地质剖面图。

此处我们看到前陆盆地的早志留世沉积层序被上部晚泥盆世含砾粗砂岩假整合 (即平行不整合)覆盖,而没有像在考察点4.4和5.1发现的奥陶系与石炭系的角度不整合。这里沉积间断的时间与点 4.4 和 5.1 相比也要短得多。

问题 1:武夷-云开造山作用在本地区有无造成地层变形?

问题 2:假整合面两侧沉积的物质成分和结构有怎样的变化?为什么?

问题 3:为什么本地区有志留纪和泥盆纪岩石的出露,而在江山地区则没有?

> **考察点 5.8 天梁公园入口处外部的晚石炭世灰岩 (28°49.017′N,118°13.132′E)**

在泥盆纪-石炭纪期间 (考察点 5.7 和本考察点),整个华南地区记录了一次海侵过程。至晚石炭世,几乎整个大陆都被台地相碳酸盐岩所覆盖 (图 1.22 和图 2.27)。

图 2.27　晚石炭世黄龙组块状灰岩

当日作业：

- 绘制本区的构造地层柱，将其与东南方仅数十千米处的江山地区作比较。

- 现在将四张构造地层柱状图整合成一个区域上的时空对比图，从左至右大致为土山、江山、双溪坞和陈蔡。识别出其异同点，并通过这张图来解读华南东部的构造演化历史。

- 以小组为单位，利用过去几天考察中看到的现象和学到的知识，以及完成的时空对比图，来准备一个有关华南东部大地构造演化历史的多媒体报告。

当日问题思考：

- 造成江山和玉山地区构造地层的显著差异的可能原因是什么？

第6日：赣东北二叠纪-三叠纪印支期造山作用的构造和地层记录，以及后造山盆地发育和岩浆活动

考察目的： 与前几天观察到的奥陶-泥盆纪间发生的造山作用及盆地与岩浆记录相对比，看该地区在二叠纪-白垩纪期间如何历史重演。

考察内容*： 应家剖面——石炭纪-二叠纪台地相碳酸盐岩逐渐被晚二叠世-三叠纪含煤线的碎屑岩（前陆盆地沉积）所取代，侏罗纪地层（造山后）与下部老地层之间的角度不整合，以及晚侏罗世-白垩纪火山岩和红层。

➤考察点 6.1　变形强烈的早中二叠世灰岩（栖霞组-茅口组）(28°14.282′N，118°00.110′E)

沿着冲沟底部行走，我们可观察中二叠世中层状灰岩，而没有造山带已靠近的特征信息（图 2.29）。山顶上同样可以看到很好的灰岩露头。在高速公路上饶入口附近(28°30.213′N，117°58.147′E) 可看到晚石炭世（或为早二叠世？）船山组块状灰岩的露头。

➤考察点 6.2　中二叠世（车头组上部）细粒砂岩 (28°14.120′N，118°1.488′E)

在通往一农户家里的土路边上出露中二叠世细粒石英砂岩。此碎屑岩的出现显示区域上沉积环境由台地碳酸盐岩相转变为前陆盆地相，表明自沿海开始的印支造山运动已开始向内陆推进（见概论第 1.7 节）。

➤考察点 6.3　位于印支造山期角度不整合面之下的强烈变形的晚二叠世（乐平组）砂岩及碳质粉砂岩和页岩 (28°14.285′N，118°2.448′E)

从图 2.28 上，你会发现此地区的侏罗纪地层"漂浮"于变形更强烈的前侏罗纪岩层之上——二者显然为不同构造期/事件的产物。

---

* 在此野外行程范围内很难寻找一条较好的线路来考察印支期（中二叠世-晚三叠世）造山作用。这些选取的考察点（图 2.28 (c)尽管涵盖了大部分的考察目标，但其露头质量良莠不齐。图 2.28 (b)中显示了一些备选点，并附上照片可与考察点做比较。

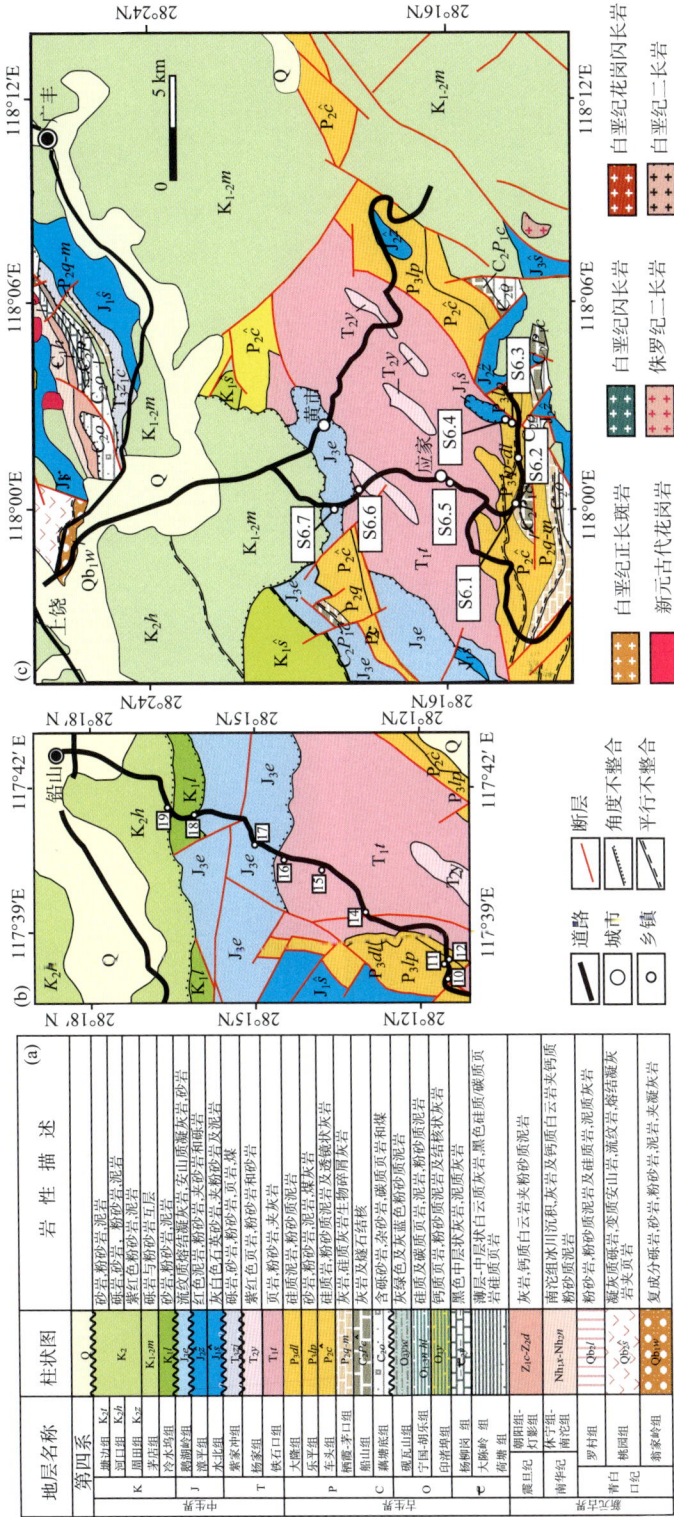

图 2.28 应豕和铝山南部地质简图以及第 6 日考察点（据 1:25 万上饶幅区域地质图改编），两幅图 (b)、(c)的具体位置见图 2.1 中的第 6 日考察位置（以⑥和⑥标出）

图 2.29 早—中二叠世船山、栖霞、茅口组灰岩

图 2.30 (a) 备选点 12 (28°11.577′N，117°38.482′E) 出露的晚二叠世乐平组中强烈变形的砂岩及黑色页岩。备选点 10 (28°11.500′N，117°38.092′E) 出露的钙质泥岩，薄层泥质灰岩和粉砂岩及其中发育的轴向劈理 (b)，以及交错纹理 (d)。备选点的位置见图 2.28 (b)

在此考察点，我们会看到晚二叠世的砂岩、碳质粉砂岩和黑色页岩出露于一角度不整合面之下，其上覆地层为早侏罗世碎屑岩 (点 6.4)。此角度不整合记录了印支造山作用对此地的重要影响。前侏罗纪地层变形明显要强烈得多——测量其层面产状，并将之与位于不整合面之上的早侏罗世地层 (下一个考察点) 的层面产状进行对比。

从二叠纪开始 (图 1.24 (b)) 到晚三叠世 (图 1.24 (d))，印支期前陆盆地从东南沿海推进到四川盆地，并在其靠造山带一侧广泛发育煤系。而在华南东南部于晚三叠世-早侏罗世发育的拗陷盆地内 (图 1.24 (d)、(e)) 也有大量煤系发育。在该区域的乐平组中有煤线发育并被当地农民采掘，但在此考察点上我们只能观察到碳质黑色页岩。

图 2.30 的照片拍自图 2.28 (b) 中标出的备选点 10 和 12，因那里地层出露更好一些。

➤考察点 6.4 位于印支造山期角度不整合面之上的早侏罗世 (水北组) 碎屑岩的构造与组分 (28°14.418′N，118°2.450′E)

在河两侧，我们可以看到恰巧位于角度不整合面之上的长石砂岩、石英砂岩 (图 2.31 (a)) 和黑色页岩。在这里将测量到的层面产状与考察点 6.3 的下伏地层产状做对比，如可能，作一表述此不整合面的信手剖面图。

图 2.31 早侏罗世水北组石英砂岩 (a) 及包含碳酸盐岩碎屑的含砾砂岩 (是砾石还是内碎屑?)

在华南的中部及东南部 (图 1.24 (d)、(e))，晚三叠世-早侏罗世地层均不整合覆盖在印支期变形地层或岩浆岩/变质岩之上。此盆地的沉积厚度可达几千米 (庞崇进等，待发表)，但却仅经历了很弱的变形过程 (通常只有地层倾斜，很少褶皱)。直到 Li 和 Li (2007) 将此盆地解释为印支造山中晚期华南东南部经历了水平俯冲及榴辉岩相变质后，由于重力拖拽而形成的大型拗陷盆地，这个不寻常的盆地才开始受到关注 (见概论第 1.7 节)。你有更好的模型来解释如此宽广 (约 1300 km) 的印支期造山带 (华南造山带) 的形成，以及发育其上的宽泛拗陷盆地吗？

> **考察点 6.5 早三叠世 (铁石口组) 粉砂岩和粉砂质泥岩的岩性及构造特征 (28°15.893′N，118°00.773′E)**

再回到印支期角度不整合面之下的强烈褶皱的二叠纪-三叠纪地层，我们在此点将会看到以薄层状钙质 (?) 粉砂岩和泥岩为特征，产状并近直立，且强烈风化的铁石口组地层 (图 2.32 (a))。在附加点 (图 2.28 (b) 中点 15)，铁石口组中发育有交错纹理 (图 2.32 (b))。在考察点更南一点的路边可以在一小处露头上看到钙质粉砂岩中发育尖棱褶皱。

图 2.32 (a) 早三叠世铁石口组近直立的钙质粉砂岩与泥岩； (b) 在附加点 15 中发育较好的交错纹理

> **考察点 6.6 位于角度不整合面之下的早三叠世 (铁石口组) 顶部发育的膝折褶皱 (28°18.368′N，118°00.407′E)**

如果此露头尚未被新建的农舍阻挡，那么这里可以看到早三叠世铁石口组浅灰色板状粉砂岩 (?) 中的膝折褶皱。要求绘制出此露头区的构造示意图，并测量膝折褶皱中的 S_0/S_1 产状。将其岩性 (反应的沉积环境) 和变形程度及样式与下一点形成于印支造山后的岩层做对比。

> **考察点 6.7 发育于印支期角度不整合之上的晚侏罗世熔结凝灰岩及其由冷凝收缩形成的柱状节理 (28°19.002′N，117°59.970′E)**

晚侏罗世地层 (鹅湖岭组) 的组分及结构与下伏地层 (我们之前观察的不整合面之下的地层) 具有明显差别。此地层主要为火山物质组成，层面平缓向北北东倾斜。此地区的火山岩及火山碎屑岩是华南中部及东南部的造山后大岩浆岩省的一部分 (图 1.25)。

图 2.33　早三叠世地层及其中发育的与层面平行的劈理及膝折褶皱

图 2.34　晚侏罗世鹅湖岭组发育柱状节理的块状熔结凝灰岩

应引起注意的是，此处火山岩的时代主要为晚侏罗世，而在更东南的地区 (第 3 日考察点) 的火山岩时代则主要为白垩纪。Zhou 和 Li (2000) 将此岩浆岩向东南方向逐渐变新的现象解释为古太平洋俯冲角度由较低角度变为高角度的结果，而 Li 和 Li (2007) 则将其解释为由俯冲板片从湘南-粤北-赣西南地区开始的拆沉过程所造成 (第 1.7 节，图 1.25) 。

当日作业：
分小组作关于本地区构造演化的多媒体报告。

当日问题思考：
印支造山作用与古太平洋的洋壳俯冲相关还是与印支地块与华南地块的陆陆碰撞相关？为什么？

浙东雁荡山白垩纪火山地质公园里的瀑布
注意地层稍有倾斜但尚未发生褶皱 (李正祥摄)

第7日： 赣东北田里片岩，新元古代裂谷岩浆作用和樟树墩蛇绿岩

考察目的： 检验更多关于扬子板块南缘在中元古代晚期至新元古代早期 (四堡造山期) 为活动陆缘边界，而在新元古代中期为陆内裂谷的证据。

考察内容： 翁家岭剖面——田里片岩及其上覆的火山碎屑岩；石龙剖面——825Ma 双峰式岩浆作用；约 1000 Ma 的樟树墩蛇绿岩。

➤ 考察点 7.1　田里片岩和角度不整合 (28°28.301′N，118°08.832′E)

田里片岩出露于广丰附近 (图 2.35 和图 2.36)，位于新元古代裂谷层序之下，是目前华南东部地区仅存的四堡造山期中级副变质岩。它可能与双溪坞岛弧系统 (图 1.1，图 1.5，图 2.35；Li et al., 2007) 一样位于同一个大陆边缘 (扬子地块南部边缘)。这里的角度不整合代表着几亿年的沉积间断，并在其上下展示变质程度上的强烈反差：未变质的新元古代中期裂谷层序 (具有底砾岩) 直接覆盖在约 1 Ga 的中级变质岩石之上 (图 2.36)。

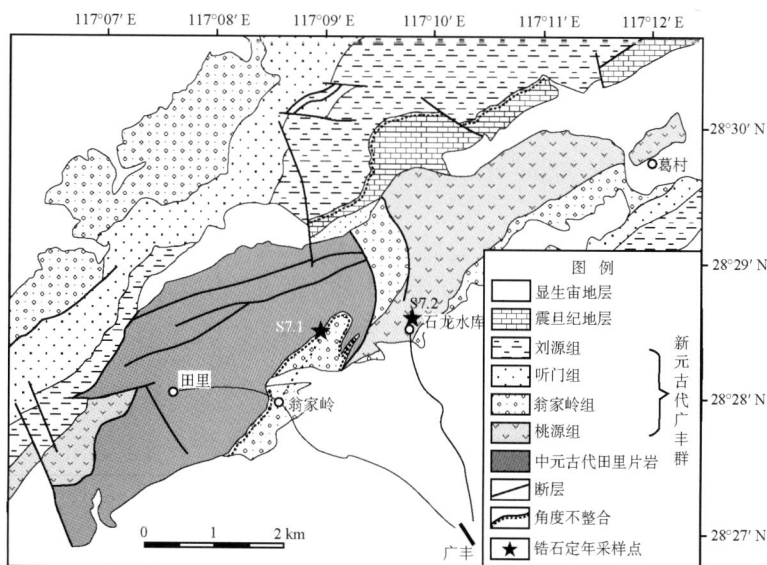

图 2.35　广丰地区地质简图 (修改自 Li et al., 2008b)

图 2.36　考察点 6.1 的简单剖面图　(据 Wang et al., 2003 修改)

　　田里片岩的锆石 U-Pb 年龄谱指示其原岩可能是源自扬子地块的碎屑沉积序列 (图 2.37; Li et al., 2007)。此原岩的沉积年龄被最年轻的碎屑锆石制约到晚于 1530 Ma，但早于其最老白云母 $^{40}Ar/^{39}Ar$ 年龄——1040 Ma (图 2.38)。如图 2.38 所示，田里片岩在四堡造山早期的 1042~1015 Ma 期间发生变质，并记录了造山晚期的 968±4 Ma 和 942±8 Ma 的构造活化事件 (图 1.5; Li et al., 2007)。

　　在此考察点我们也将观察到未变质的新元古代翁家岭组底砾岩直接沉积在田里片岩之上 (图 2.36)。翁家岭组由淡紫色火山碎屑砾岩和火山碎屑砂岩组成。砾石成分主要为流纹岩和石英及零星的玄武质砾石。而白云母片岩的碎屑则大量存在，推测其很可能来源于下部的田里片岩 (图 2.36)。翁家岭组剖面上部岩性渐变为岩屑砂岩、粉砂岩、泥质粉砂和泥岩。整体上来看，翁家岭组为一粒度向上变细的沉积序列，其中包含一些小型的正粒序旋回 (图 2.36; Wang et al., 2003)。

➤考察点 7.2　沿石龙水库堤岸公路观察约 825 Ma 的桃源组双峰式火山岩 (28°28.621′N，118°09.741′E)　和翁家岭组火山碎屑岩 (28°28.812′N，118°09.604′E ——桃源组与翁家岭组的接触部位)

　　广丰地区的新元古代裂谷层序由四个沉积相组合：①桃源组陆相双峰式火山岩；②翁家岭组冲积相和河流相的火山碎屑岩及湖泊相细粒碎屑岩；③听门组河流相长石砂岩，以及刘源组滨海相-浅海相石英砂岩；④南沱冰积岩。裂谷层序被上部台地相碳酸盐岩覆盖 (朝阳组与灯影组; Wang et al., 2003)。此剖面中我们主要观察桃源组双峰式火山岩和翁家岭组火山碎屑岩。

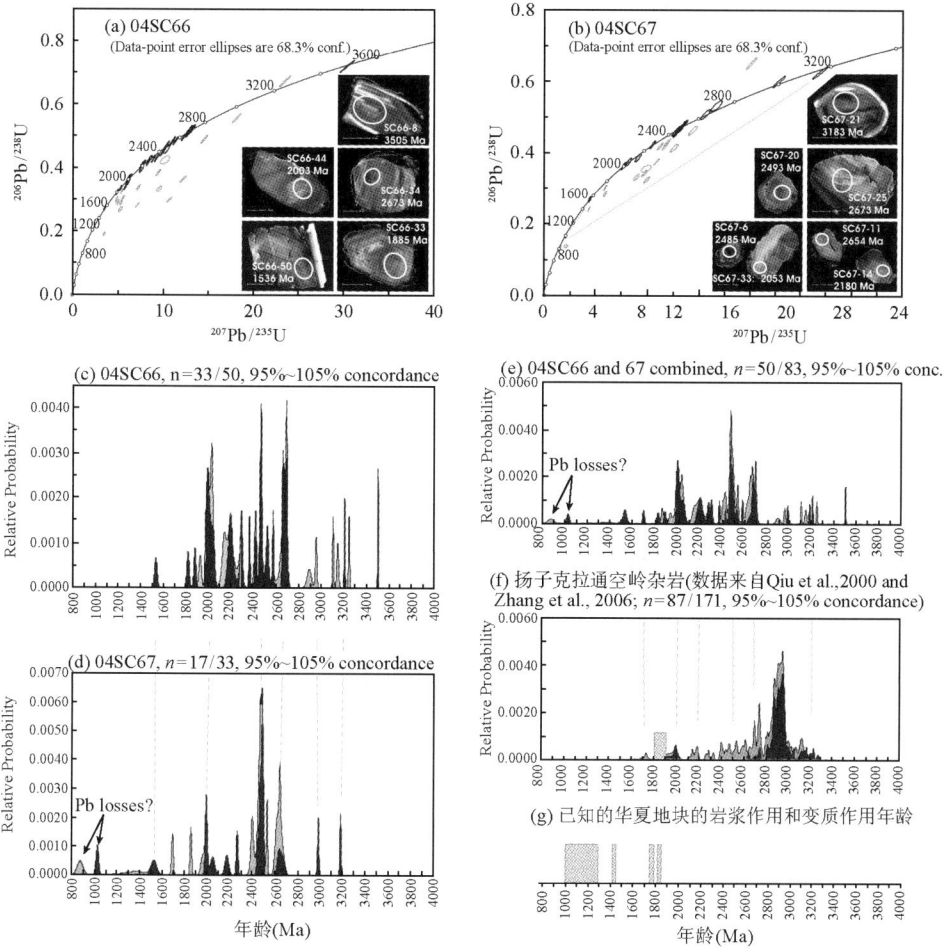

图 2.37 田里片岩的沉积碎屑锆石年龄谱 (SHRIMP) 显示其原岩具亲扬子地块的特征 (Li et al., 2007)

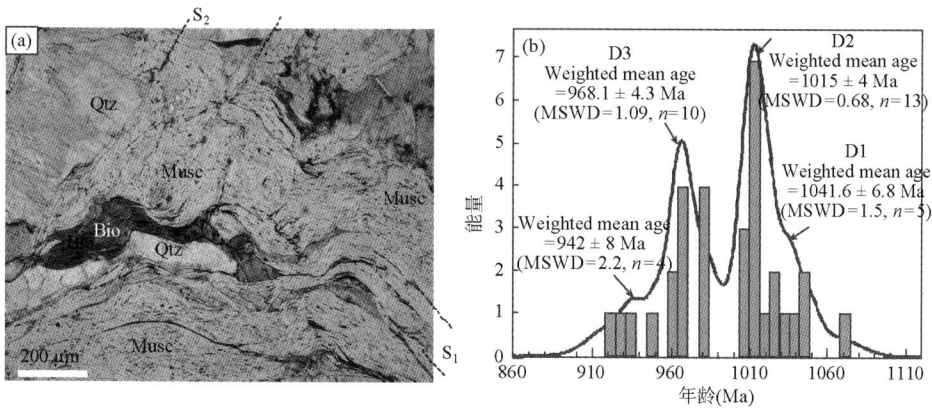

图 2.38 (a) 田里石英云母片岩样品 04SC67 中的白云母变形光学显微照片显示 S₁ 和 S₂ 片理。Musc:: 白云母, Bio：黑云母, Qtz：石英; (b) 田里片岩样品中白云母的 $^{40}Ar/^{39}Ar$ 年龄谱柱状图 (灰色柱子) 及累积概率 (黑实线) 投图 (Li et al., 2007)

桃源组总厚度约 130 m，由玄武岩、英安岩/流纹岩组成，并夹有火山碎屑岩，熔结凝灰岩 (图 2.39) 及凝灰岩。Li 等 (2008b) 对桃源组中的流纹岩锆石 U-Pb 定年结果为 827±14 Ma。而对这些火山岩的地球化学分析显示，其岩石组成为碱性玄武岩、安山岩和过铝质流纹岩 (Li et al.，2008b)。

图 2.39　石龙水库的 (a) 新元古代桃源组熔结凝灰岩的"流纹构造"，(b) 翁家岭组火山砾岩

桃源组中碱性玄武岩不仅具有 Nb-Ta 相对于 La 和 Th 的亏损，还具有高的正 Nd (t) 值 (+3.1 至 +6.0)，相对较高的 TiO_2 和 Nb 含量，高 Zr/Y 比及超球粒陨石的 Nb/Ta 比值等元素特征，显示其起源于板内裂谷环境中的，经历过板片熔融-交代变质作用的大陆岩石圈地幔 (Li et al., 2008b)。因此，广丰火山岩序列是新元古代中期板内裂谷作用过程中大陆岩石圈 (包括岩石圈地幔和中-下至上地壳) 熔融的产物 (Li et al., 2008b)。

翁家岭组假整合于桃源组之上，其厚度约 400 m，主要由火山碎屑砾岩 (图 2.39 (b)) 、含砾砂岩、泥质粉砂和泥岩组成。

更多的关于新元古代裂谷岩浆作用及可能与地幔柱关系的讨论见概论第 1.5 节及其相应的参考文献。

➤考察点 7.3　樟树墩蛇绿岩杂岩 (28°32.734′N，117°26.736′E)

赣东北蛇绿岩带 (Zhou, 1989；位置见图 1.1) 被认为是形成于扬子地块与华夏地块尚未拼合时的扬子南缘的弧后盆地或者弧间盆地 (Li et al., 1997; Li et al., 2007; Li et al., 2008a) (图 1.5)。此考察点的蛇纹岩被普遍解释为蛇绿混杂岩体的一部分，其最早的年代学结果 (1034±24 Ma Sm-Nd 矿物内部等时线年龄) 是由 Chen 等 (1991) 报道。最近，Li 等 (尚未发表) 等从此蛇绿岩中的辉长堆晶岩中获得了锆石 U-Pb SIMS 年龄 988±13 Ma，与前人的 Sm-Nd 年龄相符。

在该考察点 (图 2.40 中的 S7.3) 的岩层中发现有"枕状构造" (图 2.41)，但该构造的成因目前仍在研究中。李武显等 (工作进行中) 初步认为其岩性为异剥钙榴辉长岩，并

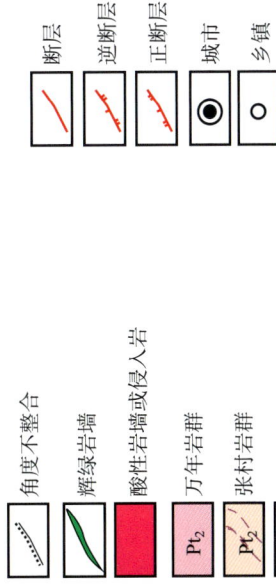

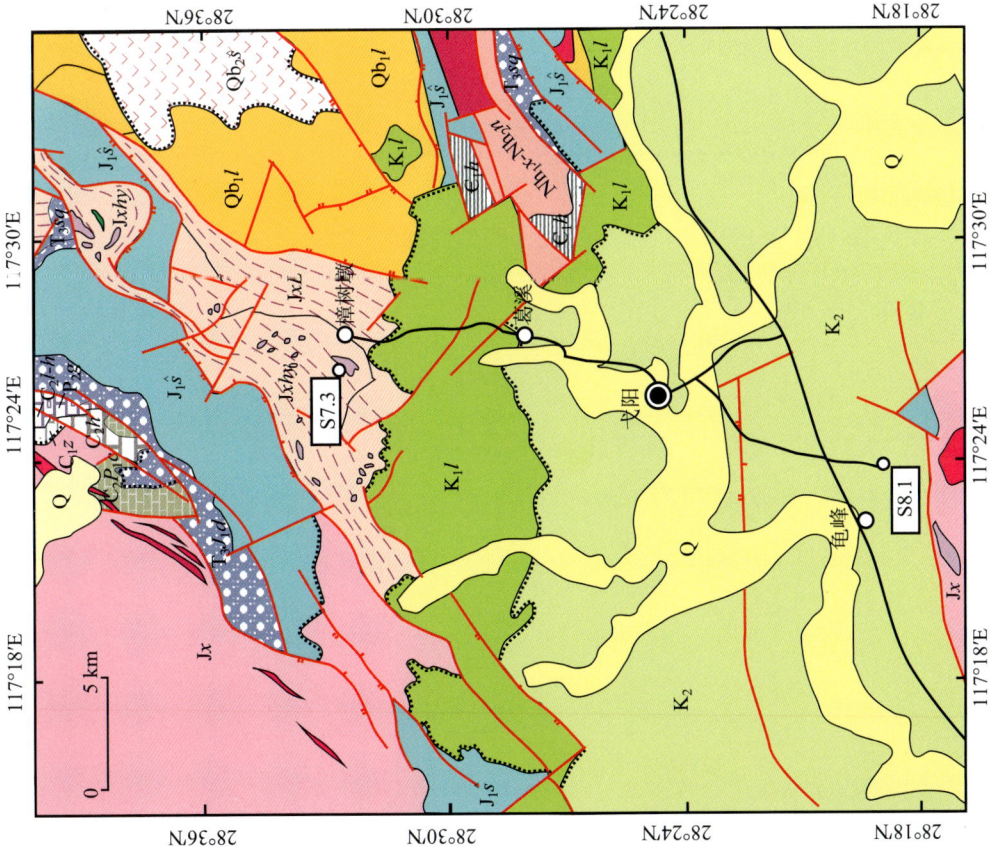

图 2.40 弋阳地区樟树墩地质图（据 1:25 万上饶幅地质图改编）

地层名称			柱状图	岩 性 描 述
第四系		Q		砂岩,粉砂岩,泥岩
K	塘边组	K_2t	K_2	砾岩,砂岩,粉砂岩,泥岩
	河口组	K_2h		紫红色粉砂岩,泥岩
	周田组	K_2z		
	茅店组		$K_{1-2}m$	砾岩夹粉砂岩
J	冷水坞组			杂色砂岩,粉砂岩,泥岩
	水北组		$J_1\hat{s}$	浅色石英砂岩及粉砂岩和泥岩
	多汇组		$J_1\hat{l}$	砾岩,砂岩,粉砂岩,泥岩
T	三丘田组		$T_1h\hat{g}$	砾岩,砂岩,粉砂岩,泥岩夹煤
	孤峰组		$P_2g\hat{s}$	硅质砾屑岩,硅质岩
P	船山组		C_2-P_1c	生物碎屑灰岩,灰岩及燧石结核
	黄龙组		C_2h	灰色灰岩,微晶灰岩
	老虎洞组-黄龙组		$C_{1l}-h$	灰色灰岩,白云岩
∈	梓山组		$C_1\hat{s}$	砾岩,砂岩,粉砂岩,泥岩,煤
	荷塘组		∈$_1h$	黑色硅质页岩,硅质岩
	朝阳组-灯影组		Z_1c-Z_2d	灰岩,钙质页岩,灰岩,白云岩夹钙质粉砂质泥岩
	休宁组-南沱组		Nh_1x-Nh_2n	南沱冰川沉积,灰岩,白云岩白云质夹钙质粉砂质泥岩
	上墅组		$Qb_2s\hat{s}$	凝灰质角砾岩,变安山玄武岩,流纹岩,熔结凝灰岩,熔结凝灰岩夹岩屑
	骆家门组		Qb_1l	暗灰、灰绿色变余岩屑砂岩,粉砂质板岩夹硅质板岩

角度不整合	辉绿岩墙	酸性岩墙或侵入岩
万年岩群 Pt_2	张村岩群 Pt_2	超镁铁质杂岩

断层	逆断层	正断层	城市	乡镇

图 2.41 樟树墩蛇绿岩中的似枕状构造

很可能形成于橄榄岩蛇纹石化过程中，由富钙和富硅的流体与镁铁质岩墙或镁铁-超镁铁质杂岩体中的其他岩石相互作用形成，而不是真正的玄武岩枕状构造。

注意此地区蛇绿岩杂岩体的"围岩"在最近的地质图中被划为"张村岩群" (图 2.40)，但这些岩石的岩性特征和年龄及变形历史仍需要进一步研究。

北东向江西蛇绿岩带已经被认为是沿着扬子板块南部边缘的弧后盆地变形 (图 1.5)。

当日问题思考：
- 樟树墩 (S7.3) 中观察的岩石代表了典型蛇绿岩套中的哪一部分？

第8日：龟峰地质公园——白垩纪红层和丹霞地貌

华南地区在白垩纪发育了大量的红色陆相盆地沉积。然而，此红层的成因仍有待研究。初步的研究表明广布的伸展或扭张盆地可能控制了红层的分布及厚度 (与北美西部的盆岭省类似)，但还需要更多更细的研究工作来清晰描绘当时究竟发生了什么。

在第8天的上午，我们将观察白垩纪沉积物组分，沉积构造及它们所形成的特殊丹霞地貌景观 (图2.40中S8.1，图2.42)。注意此处白垩纪沉积中的火山物质很少，与新昌地区所见截然不同 (第3日)：为什么？

享受下午悠闲的回家之旅！

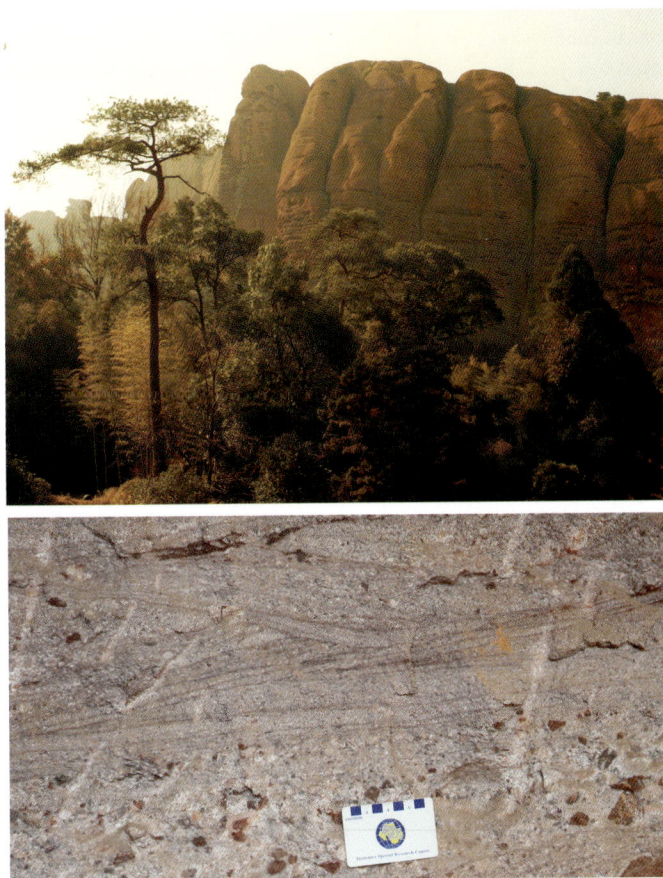

图2.42 龟峰地质公园内的晚白垩世陆相块状含砾砂岩。此套地层有轻微倾斜，但未发生褶皱变形

141

主要参考文献

董传万, 杨树锋, 唐立梅, 等. 2008. 浙江新昌复合式火成杂岩的岩石学、地球化学及其地质意义. 高校地质学报, 14 (3):365–376.

孔祥生, 李志飞, 冯长根, 等. 1995. 前寒武纪地质, 第 7 号, 浙江陈蔡地区前寒武纪地质. 北京:地质出版社:136.

马丽芳, 乔秀夫, 闵隆瑞, 范本贤, 丁孝忠. 2002. 中国地质图集. 北京: 地质出版社.

水涛. 1987. 中国东南大陆基底构造格局. 中国科学 (B 辑), 30:414–422.

章邦桐, 凌洪飞, 沈渭洲, 等. 1990. 浙江绍兴西裘双溪坞群细碧-角斑岩的 Sm-Nd 等时线年龄. 南京大学学报 (地球科学), 2:9–14.

浙江省地质矿产局. 1989. 浙江省区域地质志. 北京：地质出版社：688.

Chen J, Foland K A, Xing F, et al. 1991. Magmatism along the southeast margin of the Yangtze block:precambrian collision of the Yangtze and Cathysia block of China. Geology, 19: 815–818.

Chen Z, Xing G, Guo K, et al. 2009. Petrogenesis of keratophyes in the Pingshui Group, Zhejiang: Constraints from zircon U-Pb ages and Hfisotopes. Chinese Science Bulletin, 54:1570–1578.

Gilder S A , Keller G R, Luo M, et al. 1991. Timing and spatial distrbution of rifting in China. Tectonophysics, 197:225–243.

Eby G N. 1992. Chemical subdivision of the A-type granitoids: petrogenetic and tectonic implications. Geology, 20: 641–644.

Hsü K J, Sun S, Li J, et al. 1998. Mesozoic overthrust tectonics in south China. Geology, 16:418–421.

Jahn B M, Zhou X H, Li J L. 1990. Formation and tectonic evolution of Southeastern China and Taiwan:isotopic and geochemical constraints. Tectonophyisics, 183:145–160.

Li J . 1993. Tectionic framework and evolution of southeastern China. Journal of Southeast Asian Earth Science, 8:219–223.

Li W X, Li X H, Li Z X. 2005. Neoproterozoic bimodal magmatism in the Cathayisa Block of South China and its tectonic significance. Precambrian Research, 136:51–66.

Li W X, Li X H, Li Z X, et al. 2008a. Obduction-type granites within the NE Jiangxi Ophiolite: Implications for the final amalgamation between the Yangtze and Cathaysia Blocks. Gondwana Research, 13:288–301.

Li W X, Li X H, Li Z X. 2008b. Middle Neoproterozoic syn-rifting volcanic rocks in Guangfeng, South China:petrogenesis and tectonic significance. Geological Magazine, 145:475–489.

Li X H, Zhao J X, McCulloch MT, et al. 1997. Geochemical and Sm-Nd isotopic study of Neoproterozoic ophiolites from southeastern Chian: petrogenesis and tectonic implications. Precambrian Research, 81:129–144.

Li X H, Li W X, Li Z X., et al. 2008. 850-790 Ma bimodal volcanic and intrusive rocks in Northern Zhejiang, South China: a major episode of continental rift magmatism during the Breakup of Rodinia. Lithos, 102:341–357.

Li X H, Li W X, Li Z X, et al. 2009. Amalgamation between the Yangtze and Cathaysia Blocks in,South China:constraints from SHRIMP U-Pb zircon ages, geochemistry and Nd-Hf isotopes of the Shuangxiwu volcanic rocks. Precambrian Research, 174:117–128.

Li Z X, Li X H. 2007. Formation of the 1300-km-wide intracontinental orogen and postorogenic magmatic province in Mesozoic South Chian:A flat-slab subduction model. Geology, 35 179–182.

Li Z X, Li X H, Kinny P D, et al. 2003a. Geochronology of Neoproterozoic syn-rift magmatism in the Yangtze Craton, South China and correlations with other continents:evidence for a mantle superplume that broke up Rodinia. Precambrian Research, 122:85–109.

Li Z X, Li X H, Wang J, et al. 2003b. From Sibao Orogenesis to Nanhua Rifting:Late Precambrian Tectonic History of Eastern South China. Beijing:Geological Pubishing House.

Li Z X, Wartho J A, Occhipinti S, et al. 2007. Early history of the eastern Sibao Orogen (South China) during the assembly of Rodinia: new mica $^{40}Ar/^{39}$ Ar dating and SHRIMP U-Pb detrital zircon provenance constraints. Precambrian Research, 159:79–94.

Li Z X, Bogdanova S V, Collins A S, et al. 2008. assembly, configuration, and break-up history of Rodinia: a synthesis. Precambrian Research, 160:179–210.

Li Z X, Li X H, Wartho J A, et al. 2010. Magmatic and metamorphic events during the early Paleozoic Wuyi-Yunkai Orgeny, southeastern South China: new age constraints and pressure-temperature conditions. Geological Society of America bulletin, 122:772–793.

Li Z X, Li X H, Chung S L , et al. 2012. Magmatic switch-on and switch-off along the South China Continental margin since the Permian:Transition from an Andean-type to a Western Pacific-type plate boundary. Tectonophysics, 532-535:271–290.

Pang C J, Krapež B, Li Z X, et al. 2014. Stratigraphi evolution of a Late Triassic to Early Jurassic intracontinental basin in southeastern South China: A consequence of flat-slab subduction? Sedimentary Geology, 302:44–63.

Pearce J A. 1982. Trace element characteristics of lavas from destructive plate boundaries. *In*: Thorpe R S. Andesites. Wiley:New York:525–548.

Pearce J A. 1996. A user's guide to basalt discrimination diagrams. *In*: Wyman D A. Trace Element Geochemistry of Volcanic Rocks: Applications for Massive Sulphide Exploration Geological Association of Canada. Short Course Notes, 12: 79–113.

Pearce J A, Harris N B W, Tindle A G. 1984. Trace element discrimination diagrams for the tectonic interpretation of granitic rocks. Journal of Petrology, 25: 956–983.

Rollinson H R. 1993. Using Geochemical Data:Evaluation, Presentation, Interpretation. London: Longman Geochemistry Society: 352.

Shervais J W. 1982. Ti-V plots and the petrogenesis of modern and ophiolitic lavas. Earth and Planetary Science Letters, 59:101–118.

Vermeesch P. 2006. Tectonic discrimination diagrams revisited. Geochemistry, Geophysics, Geosystems, 7: Q06017. doi: 10.1029/2005GC001092.

Wang J, Li Z X. 2003. History of Neoproterozoic rift basins in South China:implications for Rodinia break-up. Precambrian Research, 122:141–158.

Wang J, Li Z X, Li X H. 2003. Day 3 to Day 4 morning: The Neoproterozoic rift successions and their angular unconformable contacts with the Mesoproterozoic Tianli schists (metamorphosed by the Sibao orogeny)in Guangfeng, Jiangxi. *In*: Li Z X, Wang J, Li X H, et al. From Sibao Orogenesis to Nanhua Rifting:Late Precambrian Tectonic History of Eastern South China-An Overview and Field Guide. Beijing:Geological Pubishing House:63–74.

Wang Q, Wyman D A, Li Z X, et al. 2010. Petrology, geochronology and geochemistry of ca. 780 Ma A-type granites in South China: Petrogenesis and implications for crustal. growth during the breakup of the supercontinent Rodinia Precambrian Research, 178:185–208.

Whalen J B, Currie K L, Chappell B W. 1987. A-type granites: geochemical characteristics, discrimination and petrogenesis. Contributions to Mineralogy and Petrology, 95: 405–419.

Xiao W, He H. 2005. Early Mesozoic thrust tectonics of the northwest Zhejiang region (Southeast China). GSA Bulletin, 117:945–961.

Ye M F, Li X H, Li W X, et al. 2007. SHRIMP zircon U-Pb geochronological and whole-rock geochemical evidence for an early Neoproterozoic Sibaoan magmatic arc along the southeastern margin of the Yangtze Block. Gondwana Research, 12:144–156.

Zhou G. 1989. The discovery and significance of the northeastern Jiangxi Province ophiolite (NEJXO), its metamorphic peridotite and associated high temperature-high pressure metamorphic rocks. Journal of Southeast Asian Earth Science, 3:237–247.

Zhou X , Zhu Y. 1993. Late Proterozoic collisional orogen and geosture in southeastern China: Petrological evidence. Chinese Journal of Geochemistry, 12:239–251.

Zhou X M, Li W X. 2000. Origin of Late Mesozoic igneous rocks in Southeastern China: implications for lithosphere subduction and underplating of mafic magmas. Tectonophysics, 326: 269–287.

今日引语：
你选择的并非只是一份职业，更是一种生活方式 (李正祥)

西北张家界泥盆系砂岩地貌，地层产状较缓，曾遭受印支期薄皮推覆变形